MATHEMATICS AND ITS TEACHING IN THE ASIA-PACIFIC REGION

SERIES ON MATHEMATICS EDUCATION

Series Editors: Mogens Niss *(Roskilde University, Denmark)*
Lee Peng Yee *(Nanyang Technological University, Singapore)*
Jeremy Kilpatrick *(University of Georgia, USA)*

Mathematics education is a field of active research in the past few decades. Plenty of important and valuable research results were published. The series of monographs is to capture those output in book form. The series is to serve as a record for the research done and to be used as references for further research. The themes/topics may include the new maths forms, modeling and applications, proof and proving, amongst several others.

Published

For the complete list of titles in this series, please go to
http://www.worldscientific.com/series/sme

Series on Mathematics Education Vol. **15**

MATHEMATICS AND ITS TEACHING IN THE ASIA-PACIFIC REGION

Edited by

John Mack
Sydney University, Australia

Bruce Vogeli
Columbia University, USA

 **World Scientific**

NEW JERSEY · LONDON · SINGAPORE · BEIJING · SHANGHAI · HONG KONG · TAIPEI · CHENNAI · TOKYO

Published by

World Scientific Publishing Co. Pte. Ltd.

5 Toh Tuck Link, Singapore 596224

USA office: 27 Warren Street, Suite 401-402, Hackensack, NJ 07601

UK office: 57 Shelton Street, Covent Garden, London WC2H 9HE

Library of Congress Cataloging-in-Publication Data

Names: Mack, John (John H.), editor. | Vogeli, Bruce R. (Bruce Ramon), editor.
Title: Mathematics and its teaching in the Asia-Pacific region /
   edited by: John Mack, Bruce Vogeli.
Description: New Jersey : World Scientific, [2018] |
   Series: Series on mathematics education ; volume 15
Identifiers: LCCN 2018024553 | ISBN 9789813272125 (hardcover : alk. paper)
Subjects: LCSH: Mathematics--Study and teaching--Asia. |
   Mathematics--Study and teaching--Pacific Area.
Classification: LCC QA14.A78 M38 2018 | DDC 510.71/05--dc23
LC record available at https://lccn.loc.gov/2018024553

British Library Cataloguing-in-Publication Data
A catalogue record for this book is available from the British Library.

Photo credit: Dr. Cris Edmonds-Wathen
The photo location is Biliau Elementary School, Madang Province of Papua New Guinea.

For any available supplementary material, please visit
https://www.worldscientific.com/worldscibooks/10.1142/11038#t=suppl

Printed in Singapore

Dedication

Mathematics and its Teaching in the Asia-Pacific Region is dedicated to Bruce R. Vogeli, in recognition of his lifetime of work in contributing to the improvement of the teaching of Mathematics world-wide and especially in developing nations. There are many facets of his career that reflect his activities, over more than a half-century, involving governments, school systems, field work, workshops, more than two hundred publications, conference presentations and supervision of graduate students, that amply justify this recognition. His long association with the Teachers College of Columbia University, which he joined in 1964, and where he held the Clifford Brewster Upton endowed Chair in its Department of Mathematics since 1975 until very recently, has been the lynchpin that encouraged and enabled him to bring a truly global vision into his understanding of the issues facing the teaching of Mathematics and the professional education of its teachers. This vision undoubtedly led him to involve himself with students, teachers, educators, policy developers, administrators and ministers of government in a multitude of countries around the world.

On his home campus at Columbia, he has, for example, supervised nearly 150 graduate students from developing countries through their successful master's or doctoral studies. Overseas, in addition to his own personal interactions with local mathematics teachers and educators, he has played a major role in some thirty mathematics/science workshops for

Peace Corps volunteers and host-country teachers and coordinated twenty Foreign Study tours to nations in Africa, Asia, the Pacific and elsewhere.

Dr Vogeli has had a special interest in how nations provide for the education of their mathematically talented children. He has utilised his own first-hand experiences and links with teachers in this field as the basis for a series of publications describing the various approaches that have been adopted internationally for enhancing and valuing these activities, culminating with his edited volume "Special Secondary Schools for the Mathematically Talented: An International Panorama", published in 2016 by World Scientific.

A few years ago, Dr Vogeli decided that he would complete his extensive list of authored or edited books by enlisting help from his worldwide contacts in producing a series of volumes that would provide contemporary 'snapshots' of the historic, recent and current situation in mathematics education in a number of nations around the world, with a focus on nations whose activities in this field were less well known among the international mathematics education community. This volume, entitled "Mathematics and its Teaching in the Asia-Pacific Region", is one such volume that owes its existence to his vision, to the support of the international mathematics education community and the encouragement of World Scientific.

John Mack

Contents

Contents

Preface

Dr Bruce Vogeli met with me in New York late in 2015 to brief me on his project to produce a series of 'snapshots' of the current situation with regard to mathematics and its teaching in a worldwide selection of nations. He saw this as a service to the global professional community of mathematics teachers, educators and advisors, noting that it has been a challenge for some decades now for a member of that community to gain a global perspective on this important and dynamic aspect of education. He explained his plan to devote one volume in this project to nations in the Asia-Pacific region and invited me to collaborate with him on this task. Initially, around twenty or so possible nations were considered, from mainland south and south-east Asia eastwards across the entire western Pacific Ocean. Each of us agreed to share in sounding out colleagues with regard to approaching potential lead authors for chapters.

By mid-2016, work on a number of chapters had begun, thanks to the willingness of our professional colleagues to provide good advice or accept our invitations. However, I was unaware of the steady decline in Dr Vogeli's health, which continued through 2016 and which prevented him from maintaining contact with some potential lead authors. An unfortunate consequence of this is that there are several fewer nations represented in this volume than initially planned and I truly regret my inability to redress this in a timely fashion.

The island nations represented in this volume provide the reader with an extraordinarily diverse collection of geographies, peoples and cultures, generally exhibiting two common features: a long period of colonisation or influence by the major empire-building nations of the 17th, 18th and 19th centuries, followed more recently by their emergence as independent states faced with various and often extremely difficult challenges, some inherited and ongoing, such as the changing relationships between their indigenous and imported post-colonial cultures. It is sobering to recall that world maps produced in the 19th century showed clearly the 'owners' of these now independent states and that in many cases there would have been no change even in maps published in the 1940s. Two other major challenges are the sheer number of indigenous languages in some nations or their geographical composition as a multitude of separate islands.

I offer my sincere thanks to the lead authors of these chapters, for their willingness to accept that task and for their commitment to meeting a fairly short time frame for their task. It is an additional source of satisfaction to me that they have utilised the expertise of their professional colleagues in developing their chapters. I have encouraged them to use their discretion in offering their own perspectives on the events they describe, when they found it appropriate. This may well generate discussion in the wider professional community. I stress that, as Co-Editor, I assume responsibility for the volume as published.

I am also delighted that Dr Judith Mousley accepted my invitation to her to write the Introduction to this volume, as well as to prepare its chapter on Australia with Chris Matthews.

I take an opportunity here to mention the ongoing contribution of the International Commission on Mathematical Instruction (ICMI) to the support and development of mathematics education in the Asia-Pacific region since its sponsoring of a first regional conference in Manila in 1978 and of ICME-5, the first of its International Congresses on Mathematics Education in the southern hemisphere, in Australia in 1984. The SEACME/EARCOME series of regional conferences have been important occasions for sharing of ideas and information among the nations of this region, while the staging of ICME-9 (2000) and ICME-12 (2012) in Tokyo and Seoul respectively have enabled more mathematics educators from this region to mingle with their global counterparts. The 8th EARCOME conference will be held in Taipei in 2018 and ICME-14 in Shanghai in 2020.

John Mack, Co-Editor
School of Mathematics and Statistics
University of Sydney

Acknowledgements

The critical task of identifying members of the mathematics education community in the Asia-Pacific region, who would consider accepting responsibility for preparing a chapter of this Volume, was successfully accomplished because of advice from a number of colleagues with sound knowledge of current professional colleagues with expertise in one or more of the nations included in this Volume. Those approached agreed to be lead authors for individual chapters, even though in most cases they were already heavily committed. The chapters were effectively completed in final draft form in the short time frame given for them. All of the above, and the additional co-authors of chapters, are sincerely thanked for making this Volume possible.

The follow-up work, of carefully editing each chapter and preparing the whole text for submission to the publisher, occurred in two stages. Dr Judith Mousley and Dr Max Stephens, respectively from Deakin University and the University of Melbourne, devoted considerable time assisting in reading chapters in order to identify items needing clarification by their authors and also in suggesting wordings. In addition, Dr Mousley did the final formatting prior to finished chapters being sent on to Ms Sonja Hubbert. Her expertise will convert the separate components of this Volume into a coherent whole, meeting the publisher's formatting requirements. Ms Hubbert undertook this task for previous volumes in Dr Vogeli's project and it is fortunate that she is able to execute it for this Volume.

I offer especial thanks to Mathematics Program Secretary Juliana Fullon, Professor Alexander Karp and Dr Henry Pollak, members of the Department of Mathematics at the Teachers College of Columbia University, and to Rok Ting Tan, of World Scientific, for the continuing encouragement and support they provided to Dr Vogeli and me during 2016 and 2017, as work proceeded on this Volume.

John Mack

INTRODUCTION: Mathematics and its Teaching in the Asia-Pacific Region

Judith Mousley
Deakin University

Introduction

It is an honour to be asked to write an introduction to this volume of papers about the history of mathematics education in the Asia-Pacific region. Countries represented in this volume are Australia, Indonesia, Japan, New Zealand, Papua New Guinea, the Philippines, Sri Lanka, and the South Pacific island nations.

In fact, all the nations treated in this volume are island nations. The number of constituent islands in each varies from very few to thousands; with their overall land masses varying enormously and distributed across very small, small, medium-sized, and very large islands. So geographical factors have played a significant role in the management and delivery of educational services in the past and through to today.

Each of these nations has its own historical, social, and cultural context(s) for mathematics education.

The South Pacific island nations

Andy Begg, Salanieta Bakalevu, and Robin Havea write about mathematics education in the Micronesian, Melanesian, and Polynesian islands of Fiji, the Cook Islands, Kiribati, the Marshall Islands, Nauru, Niue, Samoa, the Solomon Islands, Tokelau, Tonga, Tuvalu, and Vanuatu, focusing mainly on the twelve island nations associated with the University of the South Pacific [USP]. Between them, these islands have over 1200 indigenous languages or dialects. This area was settled by its original islanders, but was later heavily influenced by exploration, conquest, colonisation, and/or trading by Spanish, Portuguese, Dutch, British, French and German people.

As in many areas of the Pacific region, missionaries from a range of Western countries ran the first schools with the intention of "civilising" the original inhabitants and spreading Christianity. Money was introduced, and local beliefs and habits like sharing were devalued, to be replaced with Western concepts such as owning, buying, and selling. In the authors' words, "The colonists' capitalism involved land ownership, profitmaking, asset accumulation, and resource use, and those only became obvious when it was too late to return to traditional beliefs".

We learn that "until the island nations gained independence their curriculum documents, textbooks, and external examinations were usually from Australia, New Zealand, or [the relevant] colonial power (UK, France, & USA)". Even after various areas gained independence, habits of copying colonising countries were retained, and this affected curricula as well as pedagogy.

In fact, it is clear that the culture in South Pacific island natures does not meld well with Western pedagogical traditions. For example, the authors write that "group-oriented learning involves enquiry, discussion, and informal and oral group assessment" and that "group rather than individualized teaching, learning, and assessment is preferred".

New Zealand

Andy Begg, Jane McChesney, and Jyoti Jhagroo write about "colonization of land and minds" by English settlers. Prior to this, education of Māori children had been largely experiential, but the culture evidenced mathematical thinking in "counting, symmetry, measurement, patterns, buildings, and games"—although the colonisers would not have recognised the mathematics involved here. Basic arithmetic was taught in New Zealand's earliest (missionary) schools. The passing of the Education

Act 1877 established New Zealand's first secular, compulsory and free national system of primary education.

Today, New Zealand is known for high standards of statistics education in particular, as well as quality research into ethnomathematics. Begg, McChesney, and Jhagroo mention some of the people who have made significant contributions to the field. They also describe some ways that Australia and other countries have had some influence in curriculum development and discuss some mathematics education issues that remain of concern today.

I was particularly interested to read about today's Māori schools and the three Māori tertiary institutions that provide Māori-medium education for 32 000 students. New Zealand seems to be one of the few countries in the world whose native language appears to have survived colonisation, although English and Western mathematics still dominate for the majority of today's students of Māori, Pasefika, European, Asian, and other origins.

Papua New Guinea

The Independent State of Papua New Guinea [PNG] established its sovereignty in 1975 after being managed since 1884 by Germany and Britain, and then Australia. PNG is one of the most culturally diverse countries in the world, with more than 850 languages, and in many places it is extremely rugged, with some remote villages accessible only by air. Papua New Guinea comprises the eastern half of the island of New Guinea and around 600 smaller neighbouring islands. (The western half of New Guinea and its islands belong to Indonesia.) The whole of New Guinea is rich in natural resources.

The authors of the chapter on PNG are Kay Owens, Philip Clarkson, Chris Owens and Charly Muke. They write about the variety of languages across the nation; early village activities that involved mathematical thinking; the first missionary schools followed by Catholic and Lutheran influences; the earliest efforts in teacher training; the first primary curriculum; and many concerns and problems related to students having to leave their villages for schooling—some of which affect current education.

The government began schools for Europeans in the 1930s, but these were not for mixed-race, Chinese, or native children who were taught only in missionary schools. Universal primary education was first mandated in 1955, but despite government schools being free of fees nowadays, the country's current net enrolment rate of 63 per cent is the lowest in the Asia and Pacific region. Fewer girls go to school than boys, and there is a distinct divide between the elite and non-elite. Some educational control has been devolved to provinces, but the curriculum is national.

Owens and her co-authors outline how British and Australian (in particular) teachers, academics, and researchers contributed to the mathematics education scene in PNG, particularly in secondary and tertiary institutions. They also write of the impact of the Mathematics Education Research Centre, founded at the PNG University of Technology [UNITECH]. Nevertheless, we read that "from independence, the school mathematics curriculum has been little different from Australian State systems".

According to the authors, the "new math" movement was influential, along with a visit by Zoltan Dienes that "set the scene for continuing use of concrete materials". Even these days, the remote nature of many regions makes distribution of pedagogical materials and teachers' ongoing professional development difficult and the distribution of textbooks, teachers' guides, and curriculum materials challenging. Hence, activity based exercises are frequently excluded, with fairly traditional teaching being conducted in English.

Australia

The Commonwealth of Australia, one continent with associated islands, is the world's sixth largest nation by area. Early English settlers ignored the fact that its Aboriginal people had been there for (now known to be over) 60 000 years, declared it uninhabited, and claimed the country for England. Aboriginal people were recognised as citizens only about fifty years ago and as the original owners of the land only about twenty years ago. Essentially, Australia still belongs to England.

In this chapter, Judy Mousley and Chris Matthews report that the country was colonised as a penal settlement in 1788 and that "dame schools" were the first schools. They write about initial race, class, and gender inequalities and a strong "colonial legacy" that dominated mathematics education both before and after federation of the initial colonies into one nation.

The authors write about the development of secondary schools then tertiary institutions; increasing teacher education and professional development expectations; and the important roles played by professional mathematics teacher associations. Key changes in the last fifty years have included more emphasis on problem solving and modelling, the "new maths" movement, increasing participation and success in mathematics by girls, the impact of more flexible tertiary admission, and the introduction of a national curriculum. Ongoing concerns include the need for culturally inclusive education, a city-country divide, and ongoing shortages of well-qualified teachers of secondary mathematics.

Philippines

The Republic of the Philippines consists of over 7 600 islands, so its earliest peoples would have used concepts of distance and time as well as trading skills.

The authors of the chapter about the development of mathematics education in The Philippines—Ester Ogena, Marilyn Ubiña-Balagtas, and Rosemarievic Villena-Diaz—describe geometrical ideas from the prehistoric era as well as evidence of decimal and measurement use (for trading) from the country's pre-colonial times. Spanish, American, Portuguese, and Japanese colonists all played a part in the development of the Filipino education system before the nation's independence was declared in 1946.

During the Spanish era, instruction by missionaries was free and open to all. Tertiary education for sons of Spaniards and elite Filipinos began in 1611 with seminary colleges.

The first Philippine Republic was established in 1898, but this was followed by American colonisation for more than forty years. 1901 saw the beginning of teacher education. In both primary and secondary mathematics, memorization of mathematical facts and problem solving were emphasized.

The authors note that in 1941, a period of Japanese occupation commenced, and Japan tried to establish an educational system that was Asian, with mathematics being part of a production-oriented curriculum.

Independence from the USA and Japan came in 1946. Currently, the nation is involved with enhancement programs in mathematics, national assessment and system assessment programs using a reformed curriculum (K–12), and relevant professional development for all teachers.

Indonesia

Coming from Australia—a country whose population is less than 24 million—statistical data from Indonesia continually surprise me. The Republic of Indonesia is the world's largest island nation: an archipelago of over 17 000 islands with a large population of over 260 million people spread unevenly throughout the country. Indonesia has more than 50 million students and 2.6 million teachers in more than 250 000 schools.

Independence was gained in 1945 after World War II, drawing together diverse ethnic, religious, language, and school curricula. Between independence and 2016, seven different curricula have been implemented.

The chapter about Indonesian mathematics education, by Sitti Patahuddin, Stephanus Suwarsono and Rahmah Johar is organised around periods of colonial domination by a range of colonising countries— including but not limited to Portugal, The Netherlands, England and Japan—who were drawn by Indonesia's natural resources and ideal agricultural conditions as well as a wish for domination. Many of these colonising countries had significant influences on educational content and practices.

Japan

Naomichi Makinae writes mainly about modern Japanese mathematics, including some of the historical considerations that shaped the development of Japan's curriculum for mathematics. We are talked through major changes to the national curriculum across the years.

Illustrations include sample pages from textbooks, which give a sense of teaching approaches as well as content at various grade levels.

I first came across the concept of 'Jugyo-Kenkyu' (Lesson Study), which Makinae mentions frequently, through Jim Stigler whom I visited in California. Lesson study involves a team of teachers—perhaps with a mentor—working together in a cycle of observing, refining and reviewing specific lessons in order to improve their impact on student learning. Lesson study in mathematics focuses on the ideas that students need, and common patterns of students' thinking and talking about the particular concepts and their applications. Teachers discuss in detail the language, materials, and activities that can be used to convey underlying mathematical concepts. The focus in planning lessons is the consideration of the best ways for students to move from current development and expectations to where they should be at the end of the teaching period. Also discussed are the appropriate ordering of ideas and useful connections that can be made, common misconceptions and mistakes that children make, and how best to make the most of these as part of the normal teaching and learning process. Teachers raise competing ideas about such details as they study the lesson. Teachers earn recognition for publishing a description of the lesson and a summary of different opinions about points arising during their discussions. Originating in Japan, the notion of Lesson Study has received considerable international attention and adaptation.

Sri Lanka

The Democratic Socialist Republic of Sri Lanka (formerly Ceylon) is an island nation off the coast of southeastern India, with a population of over 20.8 million. Sri Lanka is ethnically, linguistically, and religiously diverse. Sinhalese people (mainly Buddhists) make up about 75% of the population.

The development of Sri Lankan education was influenced by Portuguese, Dutch, and English colonialists before its independence as the Dominion of Ceylon (later renamed the Democratic Socialist Republic of Sri Lanka) in 1948.

In Lucian Makalanda's chapter, he writes about the importance of Sri Lanka's free, compulsory education system where everyone studies mathematics for at least nine years. (Sri Lanka established free primary, secondary, and tertiary education in 1945 and is one of the few countries in the world to have maintained this.) Makalanda notes correctly that "The policies for free textbooks, meals, school uniforms and transportation, in place since the 1980s, have made Sri Lanka's education one of the most accessible in the developing world".

There are two types of government schools in Sri Lanka: national and provincial schools. These schools have different levels of resources as well as of teacher training and student performance. National schools have selective entry and a higher pass rate in Mathematics at O level. There are further differences in standards between provinces, with southern and northern areas generally having better student outcomes.

Comparing and Contrasting Information About Countries

Western pedagogies being antithetical to Indigenous culture is a theme that is common across the Asia-Pacific region. Surely, more success with educational outcomes would have come via adapting teaching and learning processes to operate within communal lifestyles and traditions. Although quite varied in their format and presentation, some of these chapters have other common themes too.

One common topic across these nations is that mathematical concepts and applications were ongoing components of their original cultures, even though imperial Western peoples rarely acknowledged these as being legitimate forms of mathematical knowledge. Even with growing acceptance of Indigenous people into societies and education systems (perhaps with children of lower status, caste or wealth), they often remained disadvantaged in formal educational opportunity. Even today, some indigenous peoples are likely to have relatively low levels of educational attainment along with higher levels of disenfranchisation. A related theme is an increasing influence—or the outright dominance—of Western mathematics.

Most of the authors write about the impact of colonial mathematics content and educational practices (and Western theorists such as Piaget and Dienes) as well as the languages, cultures, and even dominant religions of colonists—and for some countries quite a range of colonists were involved. Also notable was a common broadening of the curriculum from arithmetic to one including other forms of mathematics (including statistics as well as other branches of traditional Western mathematics).

A more positive commonality is a relatively recent shift, in theory at least, away from rote learning through a "lecturing" mode towards teaching mathematics for more active and personal conceptual understanding.

Another positive common shift was from mathematics being taught only to colonists' children in each country's earliest schools (with expectations of some students returning to the colonising country for secondary and/or tertiary education) to the development of in-country secondary schools—and eventually the development of local tertiary institutions.

The authors have all explained some common challenges for today's mathematics education and related teacher education. Most importantly, this includes challenges to traditions of pedagogical practices and content, as well as curriculum and teacher support.

A common theme that is of interest to me is the impact of changes in countries' economies over time and the effect that these had on school attendance. Some authors also note the reverse: how education (and in particular mathematics and science education) has been used as a vehicle for economic and social progress.

It is clear that the "new mathematics" movement of the 1970s affected many nations, with set theory being taught widely and emphasis being put on Boolean algebra, vectors, matrices, and group theory. Again, this was a matter of copying mother countries rather than change that brought retained benefits. Exceptions include the influence of "Realistic Mathematics Education", first developed in The Netherlands, and "Lesson Study", initially from Japan. Many of the authors also refer to the problem solving movement (and particularly its use of everyday problems relevant to students' lives); the USA-led "reform movement"; increasing use of manipulatives and "discovery learning" along with the related influence of Western theorists such as Piaget, Gattegno, and Dienes; and temporary trends such as "mastery learning" and the use of "Smart Boards".

The introduction of calculators, followed later by scientific calculators, graphics calculators, and the computer algebra system (CAS) also brought curriculum changes in many of these countries' mathematics education and assessment systems. In some countries this raised equity issues. While some countries allow the use of calculators, graphic calculators, and CAS in tests or examinations, others do not seem to. The internet is mentioned by several authors as having had an impact on the teaching and learning of mathematics—with the longer-term effects of this remaining to be seen in the future.

One thing that struck me was for how long (i.e., late into the nineteenth century) education was differentiated between indigenous students who would grow up to serve colonising countries and those who would not. This made me reflect on the self-interest of colonisers when it came to educational opportunity as well as the way such opportunity changed in many countries across the twentieth century as countries gained more independence.

A further common feature was the language of instruction, with indigenous children being expected to learn mathematics in a foreign, colonising language—although I know that this is still the case in many of the countries represented in this volume. In relation to languages of instruction, it seems that New Zealand is one of the few countries that has managed to develop schools and curricula for their indigenous students, some of whom learn mathematics (and other subjects) in Māori language. New Zealand is also experimenting with methods of teaching that best suit Māori children.

Most authors also focus attention on involvement in comparative international testing such as TIMSS and PISA, and some on the importance of international mathematics competitions.

Usually, teacher education has seen a growth in the number of years of first secondary then tertiary studies required for certification as a teacher of mathematics, although countries still differ in expectations for areas of teacher knowledge and competence.

Conclusion

In summary, these chapters provide fascinating accounts of the pressures, powers, and issues that shaped the growth of mathematics education in some countries of the Pacific and its rim. Certainly, I have learned a lot from these chapters, and I hope that other readers find them as engaging and informative as I did.

About the Author

Dr Judith Mousley taught in pre-school, primary and secondary schools for fifteen years before joining Deakin University. She coordinated mathematics education courses in the Faculty of Education's undergraduate and postgraduate programs, and had a strong record of attracting research and development funds. Her publications include edited books, chapters, research reports, journal articles, videotapes and a CD. Judy was President of the Australian Mathematical Sciences Council, Vice President (Teaching) of the Mathematics Education Research Group of Australasia, Vice President of the International Group for the Psychology of Mathematics Education, and an executive member of a range of other professional organisations. Her higher degree students worked in the areas of philosophies of mathematics education, the history of education, curriculum change, use of technologies in education, and higher education policy.

Chapter I

SOUTH PACIFIC: Mathematics Education in the South Pacific

Andy Begg
Auckland University of Technology, Auckland, New Zealand

Salanieta Bakalevu
University of the South Pacific, Suva, Fiji

Robin Havea
University of the South Pacific, Suva, Fiji

Introduction

The Pacific Ocean island nations fall into four categories.

1. Three large countries—Australia, New Zealand, and Papua New Guinea
2. Island nations associated with the University of the South Pacific/USP—Fiji, Cook Islands, Kiribati, Marshall Islands, Nauru, Niue, Samoa, Solomon Islands, Tokelau (a New Zealand dependency), Tonga, Tuvalu, and Vanuatu
3. Countries with Pacific Island territories or independent/associated countries:
   - Australia—Coral Sea Islands, Lord Howe Island, Norfolk Island, and Torres Strait Islands
   - Britain—Pitcairn Island
   - Chile—Easter Island/Rapa Nui, and Juan Fernandez Islands
   - Ecuador—Galapagos Islands

1

- France—French Polynesia, New Caledonia, and Wallis and Futuna
- Indonesia—Province of Irian Jaya (West Papua)
- Mexico—Guadalupe Island, Revilla Gigedo
- USA—Hawaii, American Samoa, Guam, Guadalupe, Northern Marianas Islands, Palau, Wake Island, and the Federated States of Micronesia (excluding the Marshall Islands)
4. Philippines—regarded as part of South-East Asia rather than the Pacific Islands

This chapter is concerned with the twelve island nations associated with the University of the South Pacific (USP). They are part of three broad Pacific Ocean regions:
1. Micronesia—(which includes Kiribati, the Marshall Islands, and Nauru)
2. Melanesia—(which includes Fiji, the Solomon Islands, and Vanuatu)
3. Polynesia—(which includes the Cook Islands, Niue, Samoa, Tokelau, Tonga and most of Tuvalu [some outer islands of Tuvalu are Melanesian])

Discovery

The Pacific Ocean covers one-third of the Earth's surface. The earliest voyagers, from Asia, Indonesia, and the West Pacific, sailed eastward and settled on the islands. Some sailed east to Rapa Nui/Easter Island, others north to Hawaii, while others reached South America and returned to the islands with the sweet potato. Such voyaging required practical knowledge of stars, ocean currents, wind patterns, and the ability to recognize the existence of islands from a distance by noting cloud formations and current changes—these 'way-finding' methods were learnt and passed on without writing or mathematics, and have been described by Lewis (1994). The people, Micronesians, Melanesians, and Polynesians, were the original people of the region and are the main inhabitants of the islands.

Spanish, Portuguese, Dutch and British ships began to sail into the Pacific in the early 1500s—some from the west and others from the east (around Cape Horn and from Peru). Later French and German ships arrived, though the German influence was reduced by their defeat in two world wars. The general intention was exploration, conquest, colonization, and trading, the exception being the Portuguese who only sought to establish trading posts. The unintended consequences of this exploration and intrusion were disease and violence.

Fiji was and is the only Pacific island nation with a significant non-indigenous racial group. Eighty-seven boatloads of labourers were indentured from India between 1879 and 1916 for the British-owned sugar-cane industry; the period of indenture ended on 1st January 1920 (Prasad, 2004); and Indians now account for about half of Fiji's population.

Tonga is the only one of these island nations that was never colonized, though they had a Treaty of Friendship with Britain from 1900 to 1970 and relied on some overseas aid, so they did experience some forms of colonial influence.

School Mathematics

Pre-colonial education

Prior to the arrival of Europeans, the islanders' languages were oral and varied across the island nations. Education was experiential and informal—elders passed on traditional knowledge acquired through experience, observation and memory. This knowledge included basic counting and ideas related to shapes, but the words used were everyday descriptors, not mathematical terms; for example, shapes were used in buildings and craft work. Pre-colonial Pacific Islanders did not partition knowledge into subjects, so the idea of mathematics as a subject, and its separation into topics (arithmetic, algebra, geometry, ...), would have seemed strange to the islanders.

Early colonial influences (1820-1949)

The initial colonizing tool in the South Pacific was education which began with mission schools. The missionaries learnt the local languages, translated the Bible, and used it as a source of stories that the local children could use to learn to read. The intention was to 'civilize' the islanders and this implied, to 'convert' them to Christianity. Having established the first schools, their emphasis was on the 4Rs (Reading, 'Riting, 'Rithmetic, and Religion [Christianity]). They taught using local languages, but with approaches based on their European experience. They assumed that civilizing implied replacing island ways of knowing and being with European ways. The traditional knowledge and beliefs of the Pacific Islanders were ignored and Christianity was assumed to provide an appropriate set of beliefs for all (even although today, these beliefs are only accepted by about one-third of the worlds' population). The missionaries were from a range of countries including England, Scotland, Ireland, France, Germany, and Holland. Their work in translating local languages was done in isolation and the island languages varied; and this variation increased during translation and while developing written forms of the languages (see appendices on pp. 33–34). Some of the schools also catered for a small number of European children.

The second colonizing tool was commerce. The colonists introduced money and a set of western commercial beliefs that differed from the islanders' traditional assumptions about sharing rather than owning, and conservation rather than exploitation. The colonists' capitalism involved land ownership, profitmaking, asset accumulation, and resource use, and those only became obvious when it was too late to return to traditional beliefs.

The third colonizing tool was language. While local languages were and are used in many primary schools, most secondary and tertiary education institutions use English. The justification for this is that there are no local words for many concepts, and that it would be neither feasible nor make economic sense to produce the required learning resources in local languages. This justification ignores two issues. Firstly, the preferred learning styles of Pacific Islanders are oral and experiential and are ignored when western reading/writing modes of education are privileged; and secondly, if languages are to survive then they need to continue to evolve by developing vocabulary to describe new phenomena.

The fourth colonizing tool (a few years later) involved making education compulsory. The educational focus was western education, western knowledge, and western concepts; the aim was to civilize which the colonists conceptualized as westernize. This resulted in the colonists' knowledge being privileged over traditional knowledge; knowing being privileged over doing, and external authority being privileged over self-control.

Over time more schools were established—church schools, private schools, and government schools. Primary schools mainly used the local language(s), and occasionally the colonists' languages—mainly English (but also French in Vanuatu, and German in Samoa). Most high schools used English. As the number of colonists increased, separate schools were set up for them, though later most schools were integrated. This integration of schools was slower in Fiji where there were local people and colonists, as well as indentured Indian labourers who had set up their own schools where teaching was in Hindi.

Late colonial influences (1950 to 1969)

Trade, religion, and colonization had led to political power being in the hands of the colonists until the 1970s when the Island nations were gaining independence. Colonization not only involved land and governance— minds were colonized too. Local beliefs were ignored as Christian, western, and commercial beliefs gained acceptance; and many islanders came to believe that European civilization was preferable to their traditional ways.

Before independence, as education became established, the teaching of all subjects, including mathematics, followed that of the colonial powers (mainly UK, Australia, and New Zealand). Primary school mathematics (for children aged 6 to 13) was limited to arithmetic as it had been for the colonists; and until the island nations gained independence their curriculum documents, textbooks, and external examinations were usually from Australia, New Zealand, or their colonial power (UK, France, or USA).

As the island nations began to gain their independence, the control of education moved to their governments, but the pattern had been set and change often involved following the example of the countries that had previously colonized them. Slowly the Departments/Ministries of Education gained more influence, though only a small number of schools were government owned—others being private or church schools. Educational budgets were limited, policy implementation was often difficult, the remoteness of many villages and islands meant education was not universally available, and high school education often required students to live away from home.

Post-colonial school influences (1970 to 1989)

Political independence was gained by all the island nations (except Tokelau) between 1962 and 1980 (see Table 1 for dates and other statistical information). In the seventies change was slow, educational funding remained limited and the reliance on policies of former colonial powers continued. Typically, years 1–8 school mathematics continued to emphasize arithmetic, and high school mathematics only changed when external examinations were changed or when new textbooks were purchased.

During this period 'modern' mathematics was introduced to schools in Australia and New Zealand; and, as high school students from the islands were sitting examinations from these countries, the islands had to update their curriculum documents and purchase new textbooks—and teachers had to adjust with very little assistance.

Because of the need to focus on education for all, some students in the senior levels of high schools, and particularly those in remote areas or on smaller islands, were not always prepared adequately for further mathematics. Consequently, pre-university mathematics courses (preliminary and foundational), equivalent to years 12 and 13 at high school, were provided at USP since its commencement, as a 'bridge' to ensure that all Pacific students have access to further mathematical study.

Table 1. Background information.

Island Nation (formerly)	Independence (from)	# Islands[1]	Land (sq km)	Population est. (Yr)
Fiji	1970 (UK)	300	18 272	881 065 (2013)
Cook Islands	1965 (NZ)	15	241	10 900 (2011)
Kiribati (Gilbert, Phoenix & Line Islands)	1979 (UK)	16	956	102 351 (2013)
Marshall Islands	1978 (USA Trust)	34	181	70 983 (2013)
Nauru	1968 (Aust/NZ/UK)	1	21	10 084 (2011)
Niue (Savage Island)	1974 (NZ)	1	260	1 190 (2014)
Samoa (Western Samoa)	1962 (NZ, was Germany)	9	2 900	190 372 (2013)
Solomon Islands (British Solomon I.)	1978 (UK)	6 (+ small)	29 785	609 883 (2014)
Tokelau (Union Islands)	— (NZ)	3	13	1 411 (2011)
Tonga (Friendly Islands)	—[2]	45 (+ uninhabited)	699	105 323 (2012)
Tuvalu (Ellice Islands)	1978 (UK)	9	26	11 992 (2013)

[1]The number of islands is approximate—there are also many rocks, atolls, islets, and reefs that are not populated.
[2]Tonga was never colonized, but was protected by a Treaty of Friendship with Britain.

Mathematics in schools after 1990

Funding for education and professional development continued to be limited in the nineties; and teaching continued to be based mainly on the teachers' prior experiences as learners—*talk and chalk*, drill and practice, textbook focus, and assessment by written examinations. This reflected the still unquestioned belief that traditional European approaches to schooling were appropriate for all learners.

The provision of education was influenced by population size, geography (number, size, and distance between the islands), and economics. Most nations had separate primary and secondary schools but many students had to live away from home to attend high school. Tertiary education was limited and often involved moving from home or learning by distance education.

Some of the current variation between islands is summarized in Table 2.

The situation slowly changed with the twelve island nations being influenced by the modern world and by the University of the South Pacific (USP) with its fourteen campuses—three in Fiji and one in each of the other eleven island nations. However, the satellite campuses are small and do not offer the full range of face-to-face programs, though some are available by distance education. Each nation now has its own government Department or Ministry of Education, most have their own pre-service teacher education institution, but USP has become the major regional influence on mathematics education.

However, in most Pacific Island classrooms the emphasis has continued to be on teachers telling (and giving notes), and students listening, reading, copying, and writing. These practices are counter to the traditional Pacific ways of learning that involved:

- oral and memory-based learning (without reading or writing),
- informal learning (by living rather than by schooling),
- experiential and practical activities (without an initial emphasis on theory),
- noticing (or becoming aware) rather than listening or reading, and
- holistic learning rather than learning partitioned into subjects.

Table 2. Some statistics.

Island Nations	No. of Islands	¹Total Population	Schools (Public & Private)			Universities	Other Tert. Insts²
			Yr1-6/8	Yr1-12	Yr7/9-12/13		
Fiji	300	893 000	721	-	172	3³	3
Cook Islands	15	21 000	25	-	5	1	1
Kiribati [2012]	16	106 000	94	9	33⁴	1	4
Marshall Islands	34	68 000	101	1	12	1	1
Nauru	1	10 000	3	6	2	1	0
Niue	1	1 000	1	-	1	1	1
Samoa	9	193 000	139	-	25	2	2
Solomon Islands⁵	6	584 000	515	154	24	1	1
Tokelau	3	1 000	-	3⁶	-	1	0
Tonga	45	106 000	126	-	50	2	10
Tuvalu	9	10 000	10	-	2	1	1
Vanuatu	80	264 000	433	-	96	1	4

Note: These statistics, from various sources between 2010 and 2015, are indicative rather than exact.
¹Population of the island nations other than Fiji is mainly indigenous, but Fiji has about 50% Indian.
²Many tertiary institutions have merged, and others may be in the process of doing so.
³USP is one of the three universities, but it has three campuses.
⁴Includes junior high schools for students of age 12 or 13.
⁵Education is not compulsory in the Solomon Islands.
⁶The Tokelau schools stop at Year 11, successful students gain a scholarship to complete years 12 and 13 in Samoa.

Mathematics in schools—now and in the future

Mathematics is compulsory in primary schools and in the first years of high school in all twelve island nations. Each nation has produced its own mathematics curriculum document and, following Australia, New Zealand, and other western countries, they all have increased the emphasis given to assessment. These initiatives began with little or no consideration being given to local needs, ways of coming to know, or the realization that assessment gives students the message that they are not succeeding.

Some mathematics educators are concerned because some indigenous students are not using deductive or developmental thinking and wonder, how might this situation be changed? Their concerns and responses are based on western views in which traditional ways of 'thinking, knowing, learning and being' of the indigenous people in the Pacific are not acknowledged—western ways are taken for granted and assumed to be superior. The alternative response would be to ask, 'how do indigenous people learn, how might we cultivate these ways, and can western people learn from these Pacific ways of learning?'

A related concern is that the majority of Pacific islanders who influence mathematics education policy have been educated in systems dominated by European thinking. This has influenced them and as a consequence they take European ways for granted and assume that others can adapt as they did—their minds have been colonised by western education. As a consequence, many people in positions to facilitate change assume that the need is to 'improve in western ways' rather than to 'change in ways that fit with their own cultures.' However, it is not one approach, or another, but both—our world is multicultural and multicultural solutions need to be found to questions such as:

- Do alternative curriculum foci exist, what are they, and how might they be realized?
- Would a curriculum that integrated subjects be more appropriate for Pacific students?
- Are there multicultural approaches that fit with western and indigenous thinking?
- How might European mathematical concepts be taught in multicultural ways?

- Should more emphasis be given to visual rather than symbolic argument and proof? For example, when proving that $(a + b)^2 = a^2 + 2ab + b^2$.
- How might more emphasis be given to oral/aural rather than writing/reading approaches and how might these be grouped rather than individually focussed?
- How might mathematics education change so that it fosters imagining, visualising, using intuition, developing awareness, and using logic?
- Most mathematics teachers assume that mathematics is 'true', but truth is relative to assumptions, and students may not be aware of this, (e.g., $7 + 9 = 16$, or 4 on the clock).

Educational Development

The four main forms of educational development are teacher, curriculum, resource, and assessment development.

Teacher development

Teacher development includes pre-service education (at teachers' colleges, USP, and other institutions), and in-service education.

Pre-service education in the past was predominantly western and did not adequately consider Pacific ways of teaching, learning, knowing, and assessing; however, this is slowly changing. In-service education should occur in all schools at staff meetings, but often these meetings are administrative rather than professional and teachers with other commitments are not able to attend.

In-service education can occur at professional teacher meetings, but this is only practical in larger towns or where schools are in close proximity. Radio and television offer two alternative modes for teacher development, but typically they are used for administrative instructions rather than development. (Administrators say 'what to do', while developers ask 'What should we do?') The most successful development occurs when colleagues work together in the same school or in nearby schools.

If mathematics is to be learnt by Pacific Island students, then teachers need to adjust their teaching approaches to ensure that they are culturally responsive (rather than following western practice). To do this they need to first identify the practices that are culturally responsive. These practices include group work, building on prior experiences, oral and practical tasks, and so on. These alternatives are fundamental and challenging, and teachers may require professional development if changes are to be made.

While USP is the main regional tertiary education provider, other providers need to be acknowledged. These include the other tertiary education institutes, teacher and subject associations, school-based discussion groups; and more recently, professional reading facilitated by the increased availability of online resources, and MOOCS (though online resources may need to be modified to fit within a Pacific context). Supplementing these are school staff meetings (and staff and parents' meetings) that focus on professional learning rather than on reporting or administration.

Development is a shared responsibility of teachers, schools, Ministries of Education, and tertiary institutions, but starter resources may be needed to assist schools (individually or in clusters), subject or teacher organizations, and individual teachers to initiate professional development activities. However, teacher development is only useful if it actually results in change in practices, beliefs and attitudes.

One further difficulty is ensuring that all teachers have had adequate teacher education—some are still untrained. In some remote areas schools rely on untrained teachers, but they would be understaffed if all teachers had to be trained. Additionally, some of these teachers have family commitments that mean it would be difficult for them to leave their homes to complete a teaching qualification.

Curriculum development

All Pacific Island nations value their independence, but most have curriculum documents based on western models with knowledge partitioned into subjects and topics and not considered holistically, and, for many teachers, the textbook is the curriculum. The partitioning of subjects (or textbook chapters) into topics and lessons reflects the view of curriculum as the 'tree of knowledge' (with branches representing subjects, twigs representing topics, and leaves representing lessons). An alternative and holistic view is to think of the curriculum as 'rhizomatic' (a rhizome being a creeper that grows under the ground, from which apparently disconnected shoots seem to appear randomly, yet are connected by the plants roots). This rhizome metaphor implies the teaching of themes rather than subjects and reinforces the connected nature of all knowledge that fits with indigenous Pacific ways of knowing.

If the curriculum is intended to be *'all that is planned for the classroom'* then it needs to involve: aims, learning, teaching, assessment, and contexts as well as content, with aims being the most important. However, the main focus of Pacific curriculum documents, particularly at high school level, is *'separate subject content knowledge to be taught and assessed'*. This separation based on western ideas of schooling together with the current emphasis on assessment leads to teachers 'teaching to the test'. Consequently, most curriculum documents emphasize content and ignore learning and teaching approaches.

Typically, the island nations have developed their curriculum documents by looking to Australia, New Zealand and other western countries for guidance rather than to their own cultures for content, concepts, contexts, and teaching approaches. It is interesting to consider what a Pacific Island curriculum might look like if the developers allowed their cultures to influence their curriculum documents.

Traditionally, school mathematics curriculum documents have listed numerous disconnected topics and subtopics to be taught and assessed; but other possibilities exist that may be more appropriate, these include:

- numeracy (i.e., emphasizing arithmetic skills),
- problem solving and problem posing,
- topic integration (for example, using 'relations' as a unifying notion),
- technology (calculator/computer) use, and
- cross-curriculum tasks that do not separate knowledge into subjects.

Pacific nations have much to consider as they move forward. One challenge is to have teachers interpret rather than follow the curriculum or textbook (a commercial curriculum), and a second is to ensure that the emphasis on preparing students for examinations does not replace developing broad and contextual understanding. Unfortunately, Pacific curriculum developers have focused on western ideas about mathematics rather than on island ways of knowing, learning, teaching, and assessing. This has resulted in an emphasis on: listening, reading, and writing; rather than questioning, discussing, observing, exploring, and connecting learning with living.

This situation may slowly change with computers (and the one-laptop-per-child initiative has helped this) but this long-term influence depends on whether computers continue to be distributed and whether western (English language) resources or local culturally appropriate resources are created and shared.

Resource development

Related to curriculum development is resource development; and this is important as resources often become the de-facto curriculum. Pacific Island resource development for the first few years of school is local (and local languages are used), but the resources are often translations of English language resources with minor modifications; they relate to specific subject topics rather than to areas of student interest, and they imply a reading/writing (western) orientation, not an oral/aural (Pacific) one. Additionally, high school resources are typically in English and are subject specific rather than integrating knowledge in traditional Pacific ways.

An emerging resource is the world-wide-web and associated technology. While this has a western rather than a Pacific flavor, it is increasingly available at low cost. Web-based resources cannot be ignored, but care is needed to ensure a balance between indigenous and western education. Additionally, aid money is sometimes available to purchase overseas resources—but these rarely use Pacific contexts or focus on holistic knowing. The use of the web and overseas resources results in western thinking dominating, colonizing, and belittling traditional ways of knowing.

Assessment development

The emphasis on western assessment (formal, written, subject-specific, and individual), contrasts with what might better suit Pacific students (informal, oral, holistic, diagnostic/formative, groups and self-assessment). Informal assessment (listening to students, talking with them, and asking questions) need not involve tests that 'teach' students that they cannot do mathematics. Changing from formal written assessment to informal in-class group discussions is more culturally appropriate and feeds back into learning. When written assessment is necessary, there is a need to remember that low marks and crosses on pages are a disincentive—constructive comments rather than marks or grades are more appropriate.

There are three main purposes for assessment—diagnostic, formative and summative, (before, during, and after teaching); but, with any topic, students always have some prior ideas, and there is always more to learn about the topic, thus, all assessment is essentially formative. The need for summative assessment needs to be interrogated—assessment is usually about what has been taught, not what was learnt; in spite of the relevance or irrelevance of the topic; and when grades are low the student is blamed rather than the curriculum or the teacher.

The old adage 'no one gets taller by being measured more often' is relevant when considering assessment—marks do not encourage learning, and negative messages lead to students giving up. Unfortunately, these effects are often ignored as is the notion that a grade is not as helpful as a constructive comment.

It is often assumed that assessment stimulates improvement, but it also narrows what is taught to teaching 'to-the-test'. Fasi (2005) has emphasized the importance of assessment but also acknowledged that performance in high-stakes assessment should not be taken as an indication of quality, and this together with the merging of subjects and the increasing use of computers, is causing a questioning of the place of assessment. He has suggested the possible need to shift to self-assessment, and to modify tertiary education entry requirements to allow more democratic entry.

Financial challenges

Currently educational development in the island nations is initiated and undertaken independently by each country, but the nations' educational budgets are limited. From time to time financial assistance is available from external sources (Australia, New Zealand, or aid agencies), however, such reliance is colonialism if the ideas from the funding source take precedence over local ideas and needs. Aid agencies and donor nations are modern colonists rather than philanthropists if their agendas align with the thinking of their countries rather than that of the recipient island nations.

One reason for the acceptance of this ongoing 'colonizing' (or 'post-colonial westernizing') is that many Pacific Island educational policy-makers are products of the colonial systems that existed in the islands when they were educated. They believe that their education fitted them well for their current decision-making roles; they are grateful for the education they had and what it offered, and most question policy decisions from an international rather than a local position. However, in the last few years a number of indigenous educators have questioned the western colonization of education, but they are pragmatic. They know that one cannot turn the clock backwards but they believe that there is a need to recover their cultural heritage which includes aspects of past pedagogical practice and traditional knowledge.

Influences on School Mathematics

Eight influences on school mathematics are: educational aims, language, other subjects, ways of knowing, cultural issues, gender, finance, and ongoing colonization.

Educational aims

Educational aims should be the major influence on educational practice, but most curriculum documents emphasize assessable subject content knowledge rather than aims, so teachers focus on subject content knowledge. We need to ask whether the aim of education is knowledge accumulation, and if so, whose knowledge? Alternatively, is this aim for qualifications, employment, or life, and with what implications?

Encouraging thinking is an educational aim (though emphasized differently in different societies). Mathematics teachers emphasize logical thinking, but other forms of thinking are also relevant to the learning of mathematics (and other subjects)—these include: empirical (sensory/experience-based) thinking, creative thinking, meta-cognitive thinking, caring thinking, cultural thinking, contemplative thinking, subconscious thinking, and systems thinking. Using all these forms of thinking broadens one's view of mathematics.

Language

The Pacific region has over 1200 indigenous languages or dialects, and the nations linked with USP use nearly 500 of these; see Table 3.

The language of instruction in primary-schools is normally the local language—the exceptions being when many students are English, Indian (in Fiji), French (in parts of Vanuatu); or when there are many different indigenous languages and the decision is made to use Pijin (in the Solomon Islands) or Bislama (in some schools in Vanuatu). The appendix at the end

Table 3. Languages of the South Pacific.

Pacific Island Nation	Local Languages	Dialects	Colonists' Language(s)	Other Languages
Fiji	Cook Is Māori	8	English	
Cook Islands	Fijian	300	English	Hindi (var.) Pidgin Fijian
Kiribati	Kiribati	2	English	
Marshall Islands	Marshallese	2	English	
Nauru	Nauruan		English	Nauruan Pidgin
Niue	Niuean	2	English	
Samoa	Samoan[1]		English	
Solomon Islands	60 to 70 languages		English	Pijin
Tokelau	Tokelaun		English	
Tonga	Tongan		English	
Tuvalu	Tuvaluan	2 groups	English	
Vanuatu	103 local languages		English, French	Bislama

[1]Variations of the Samoan language were used when speaking to chiefs.

of this chapter (pp. 33–34) shows some of the differences between the local languages, and these differences make resource sharing impractical.

However, high school teachers and tertiary lecturers use English because:

- it is a common shared language of many of the students,
- no local translations exist for most of the 800 mathematical terms,
- textbooks are published in English (and it is not practical to translate them),
- numerous lecturers and teachers do not speak the local languages,
- external high school examinations use English, and
- high school students need to master English for work and tertiary study.

Additionally, there is little recognition that the language of textbooks and of mathematics teachers and lecturers differs from everyday language and this can confuse students.

Other subjects

The separation of mathematics from other subjects and the partitioning of mathematics into topics (arithmetic, algebra, geometry, statistics, …) are western cultural constructs that differ from the holistic way that many indigenous Pacific Island people conceptualize knowledge. Thus, to teach mathematics as a separate subject is a form of colonization; and this leads us to ask how teachers might integrate the learning of subjects. The western compartmentalization of knowledge into subjects is currently being challenged by Finland (the European leader in education) which is introducing a curriculum without subjects—perhaps Pacific countries will resonate with this notion in the future.

One way in which mathematics links with other subjects is through the use of application-based problems. However, when resources from developed countries are used, the applications usually relate to the source country rather than to the Pacific. Contexts for applications need to be local and practical, interest the students, and include ideas that are often not associated with mathematics. These might include:

- environmental issues (renewable resources, coastal erosion, hurricane-proof building, …),
- food (farming, fishing, fruit and vegetable gardening, cooking, self-sufficiency, …),
- transport (coastal navigation, transport timetables, alternative transport systems, …),
- sport (individual/teams, competitive/non-competitive, organizing sports draws, marking fields for athletics, rugby, netball, …),
- traditional and modern arts and crafts (carving, weaving, building, sewing, …), and
- other school subjects and everyday leisure activities.

Too often western educators assume 'teach the concept first, then apply it'; yet mathematical applications are often more appropriately used when a topic is introduced. This is because the students are more likely to be familiar with the application and it provides 'concrete' meaning to a lesson and a practical situation to scaffold theory.

Ways of knowing and coming to know

There are numerous differences between western and indigenous ways of working (see Table 4) that are part of the 'hidden' curriculum. They need to be considered as they can result in different understandings.

These ways of working imply a need to reconsider mathematics education in the Pacific by asking:

- should teachers use or change their students' ways of working?
- how can teachers respond to students' preferred ways of learning?
- is the teachers' role to instruct, to explain, to encourage exploration, or to let learn?
- should mathematics be considered as a subject or as part of holistic knowledge?
- could assessment be less formal and involve cooperative, group, or self-assessment?
- is learning about making connections, remembering, or understanding?
- how might group work rather than individual work be encouraged?
- can visual, oral, movement and gestural modes of learning, rather than verbal modes (listening, reading and writing) be encouraged? and
- what familiar cultural contexts that relate to student life can be used in classrooms?

Two further differences that need to be considered when planning education are: in the west the curriculum is organized into separate subjects, and classes consist of students of similar ages—but Pacific Islanders traditionally conceptualized knowledge holistically; and education involved broad age groupings. These differences have implications for curriculum documents, and school/class/lesson organization. A third difference is that learning was experiential and traditionally achieved by doing, it did not involve formal tuition or the separation of knowing into subjects. Learning emerged from playing, caring, cooking, gardening, canoeing, fishing, building, and so on, and these tasks were often interconnected. Only a few children (mainly high-

Table 4. Western and indigenous ways of working.

FIELD	WESTERN WAYS	INDIGENOUS WAYS
Knowledge	Partitioned (into subjects)	Holistic (connected
Educating	(Individually)	(Group-based)
Teaching	Telling/explaining/showing	Asking/experiencing/modelling
Learning	Listening/reading/writing	Trying/discussing/observing/
Thinking	Logical/creative	living
Knowing	Knowing-of	Intuitive/empirical (sensory-
Assessing	Achieving (summative)	based)
Communicating	Controlling	Knowing-how
(Dominant modes)	(Reading/writing/verbalising)	Becoming aware (formative
		Noticing/connecting
		(Listening/talking/picturing)
Mathematics	Formal logical (proof)	Informal intuitive (evidence)
	Manipulating	Modelling
Science/Technology	Experimenting/exploiting	Explaining/conserving
Commerce	Consuming/Profiting	Conserving/preserving
Social Studies	Land-based/written/power	Island-based/oral/people (family)
Philosophy	(Intellect-based)	(Experience-based)
Ontology (being)	Individual	Family/humanity/environment
Epistemology (knowing)	Individual knowing	Relational (connected) knowing
Believing (religion)	Restricting (authoritarian)	Revealing (experiential)

born boys), were taken aside and formally taught (in preparation for leadership roles that they might inherit).

Learning contexts

In terms of learning contexts for schools and tertiary-level education, it seems appropriate to consider the Declaration of Barbados (United Nations, 1994) from the United Nations Global Conference on the Sustainable Development of Small Island Developing States. This noted that many small developing states are vulnerable to natural, environmental, and man-made disasters such as climate change (global warming), weather (hurricanes), rising sea levels, tidal waves, and erosion—and that these impact on: land, housing, water and food supplies, and the survival and safety of the population. Such disasters, whether natural or man-made, imply a need to develop sustainable and self-sufficient ways of living, which imply a need to refocus education (and the curriculum) by shifting from subject knowledge, to 'practical and useful' knowledge. Such a curriculum would not only be mathematically and statistically rich, but also relevant to learners whether the disasters occurred or not.

Cultural considerations

According to D'Ambrosio (2001), ethnomathematics is about the relationship between culture and mathematics; it involves using cultural artefacts as sources of contexts and applications for problems in mathematics. Some examples that have been used in geometry include: designs on tapa cloth, mats, fabrics and carvings; and the shapes of traditional buildings; but very little has occurred in the other areas of mathematics.

However, care is needed. Craftspeople who create artifacts are often unaware of the possible mathematical interpretations and may feel that their status is diminished by the mathematical discussion; thus, teachers need to ensure that they are adding a dimension to the artifacts and to the 'mana' of the creative craft workers.

Apart from contexts, Pacific Islanders traditionally learnt many things, but not by formal or logical instruction, so changing the teaching and learning modes may well be more important than the applications.

Thaman has been writing about indigenous knowledge and culture for many years (see Thaman, 2001, 2009, 2013). In her 2013 paper she specifically addressed the need for teachers to be both 'professionally qualified' and 'culturally competent'. Thaman emphasised that this was particularly important for teachers of indigenous students who face conflicting expectations from schools and homes, and that the challenge in teacher education is, "to prepare people to live in an increasingly globalised world and at the same time, develop systems of education that will ensure the survival and continuity of their (Pacific) cultures" (p. 99).

Her writing implies that those involved in Pacific education need to be aware of the values and knowledge systems of their cultures, and how their cultures are evolving.

Thaman's concern about the mismatch between western-style and Pacific Island education, and the need for Pacific educators to take into account indigenous knowledge and culture, has also been noted by other authors. For example, Bakalevu, Tekaira, Finau, and Kupferman (2007, p. 70) identified the need to "incorporate cultural elements in the formal curriculum, and to use culture-sensitive pedagogies". Such pedagogies need to include:

- informal-group rather than formal-individual approaches to learning and assessment,
- oral and experiential approaches rather than written and textbook-focussed work,
- local rather than western language use (including mathematical terms),
- creative and intuitive rather than logic-based thinking and learning, and
- broad/holistic (multi-discipline) topics rather than specific learning objectives.

If mathematics and mathematics education are to be more culturally acceptable then cultural applications is a start, but this needs to be made explicit in curriculum documents, teachers' guides, and textbooks; and the influence of culture on teaching and learning also needs to be explicit. Three related cultural issues are:

- mathematics is part of a western partitioning of knowledge, but Pacific knowledge is holistic, thus an integrated curriculum may be more appropriate;
- group rather than individualized teaching, learning, and assessment is preferred; and
- Pacific Islanders learn experientially rather than logically with hands-on activities and with time for ideas to come together naturally rather than through formal teaching.

In European education, mathematical topics typically come directly from a textbook (that is often the de-facto curriculum) or from a formal lecture with the lessons being teacher- or lecturer-directed. Learning is reinforced by students working individually and completing examples from the textbook; and formally assessed by individuals answering assessment tasks. This is the antithesis of the traditional Pacific education where topics arise from the students' experiences and their environment; and group-oriented learning involves enquiry, discussion, and informal and oral group assessment.

The mismatch between practice and the Pacific ways of working imply a need for change with respect to many aspects of educational organization and 'delivery' in the island nations. This involves curriculum reorganization, appropriate resources being available, and changing teachers' practices from a delivery mode to a guidance mode. However, whether this can be prioritized is problematic—it would require professional development and there are many competing demands on the educational budget.

Gender issues

The international challenge of girls not thinking that mathematics is important has been recognized around the world for some years, and support groups have been set up to help teachers address this issue in many countries. This issue is evident in the Islands; girls rarely consider STEM (Science/Technology/Engineering/Mathematics) based careers. But formal and informal mathematics are needed for more than STEM-based careers, they are required in: design, architecture, accounting, business, economics, social sciences, non-professional careers, and day-to-day living. Teachers need to use a range of applications so that all students become aware of the value of STEM subjects (rather than focus on STEM careers that may be irrelevant in the islands).

Tertiary Education Providers

Tertiary mathematics is typically vocational or academic. Vocational mathematics has been taught in the islands as needed by the various tertiary institutions that predated the University of the South Pacific (USP) that was established in 1968 with its main campus at Suva in Fiji, and smaller campuses on each of the twelve island nations. (The smaller campuses do not offer the full range of courses, but they do sometimes host tutorials to support distance courses.) USP was modeled on English universities and is the main academic (and vocational) tertiary institution in the region. Some of the tertiary institutions that predated USP remain, but many have merged.

Mathematics at USP

Pre-degree mathematics

USP was established in 1968 with three schools: Education, Social Development, and Natural Resources. Mathematics was among the first disciplines to be offered and this began in the School of Education with four mathematicians teaching the subject. The initial focus of the School of Education was teacher preparation and as part of this papers (units of study) in mathematics were offered at the pre-degree level to ensure that future teachers could teach mathematics at all levels. This was important because numerous high schools in remote areas did not have specialist mathematics teachers and many did not have Form 6 and 7 (years 12 and 13) classes.

USP's College of Foundation Studies

The pre-degree papers that were initially taught by the School of Education are now offered by the USP's College of Foundation Studies at two levels—the preliminary level (equivalent to High School Year 12/Form 6) and the foundation level (equivalent to High School Year 13/Form 7). The college offers about 50 papers, seven focus on mathematics and statistics, three on computing and information systems, and others relating to language, communication, sciences, social sciences, commerce, and economics. Currently many local students take these pre-degree papers rather than attend their local high schools, and some students from other islands take them because education in their home islands is limited. This college is well staffed with nearly 50 people in Suva and 20 others on five other campuses.

Degree-level mathematics

By 1970 six degree-level papers covering algebra, statistics, calculus and computing were offered by mathematics lecturers in the School of Education. As the University grew it became necessary to form departments and in 1977 a separate mathematics department emerged with Mr. Ross Renner, a New Zealander, leading the group. In 1980 this group became a full Department of Mathematics; and from 1980 to 1983 it was led by Dr. Donald Joyce from New Zealand who was a Reader of Mathematics (Reader being equivalent to Associate Professor).

From 1985 to 1987 Professor Ted Phythian from the UK led the department and in 1986 it became part of a newly formed School of Pure and Applied Sciences which had formerly been the School of Natural Resources. Professor Donald Richards from the UK was the next head of department from 1989 to 1994 and during his time in 1990, the Department of Mathematics was renamed the Department of Mathematics and Computing Science to reflect the growing importance of information technology. In 1990, 11 mathematics and 7 computing science papers were offered, and by 1993 the number had grown to 12 in mathematics and 9 related to computing—these all being of one-semester duration.

In 1995 and 1996 Professor Ted Buchwald from the UK led the department. He was followed by Associate Professor John Hosack from the USA from 1997 to 2003, whose interests were both mathematics and computer science; and by the end of 2004, there were 17 computer science/information systems papers and 18 mathematics papers offered at the undergraduate level. Next, Associate Professor Jito Vanualailai, a mathematician, led the school from 2004 through to 2009 and during this period a Faculty-based structure was established at USP with the Department of Mathematics and Computing Science being transformed in 2006 into the School of Computing, Information & Mathematical Sciences with responsibility for four disciplines—computing science, information systems, mathematics, and statistics; and with the school sitting within the Faculty of Science, Technology and Environment. Professor Eduard Babulak, a computer scientist, led the school from 2010 to 2011; then late in 2011 Dr. MGM Khan was appointed acting Head of School until Professor Ansgar Fehnkar, a computer science specialist, took over the leadership role from 2012 to 2015. Associate Professor Khan was again appointed as acting head of school in 2016 until Professor Peter Croll, a computer scientist from Australia, arrived late that year to take up the leadership role.

In 2016 the school offered 63 papers at the undergraduate level, these comprised: 13 mathematics papers, 2 statistics papers; 24 computer and information science papers, and 24 engineering mathematics papers. For Masters degrees, the courses offered were 4 mathematics papers, 3 statistics papers, and 12 computer and information science papers. Over the last few years, financial pressures and falling numbers of enrolments have caused some undergraduate and graduate mathematics and statistics courses to be discontinued, though computing has not been affected. Acknowledging that change is inevitable and with the increasing demand for computing and the ability of computers to do many tasks in computational ways that do not require calculus, one wonders, how might the demand for mathematics change in the future?

Mathematics Education

USP

Mathematics education papers (as distinct from mathematics and associated topic papers) are concerned with the teaching and learning of mathematics rather than the mathematical content. At USP, mathematics education papers are taught by education department lecturers as 'teaching methods' papers for the BEd degree, which is USP's qualification to teach in a primary school in the South Pacific region. Mathematics education involves three papers, one at each year level. In addition, within the BEd there are other papers related to teaching, assessment and evaluation, curriculum studies, curriculum development, and assessment that are relevant to all subjects including mathematics.

Prospective high school teachers typically need to complete a four-year course that involves a bachelors-level degree (usually a BA, BSc, or BCom) which covers the two teaching subjects that they expect to teach at high school level; together with a Graduate Certificate in Education. These qualifications qualify them to teach in high schools in the South Pacific region. The courses for the combined qualification include:

- eleven education papers including a practicum course,
- seven or eight papers in the first discipline major that they expect to teach,
- seven or eight papers in their second discipline major, and
- sufficient electives to make up the required thirty-one papers.

While the education courses are offered by the School of Education which is in the Faculty of Arts, Law and Education, the papers involved in the two discipline majors are taught by USP staff outside of this faculty. Mathematics, related subjects (computing science, information systems, and statistics) and other science subjects, are offered by the School of Computing, Information and Mathematical Sciences within the Faculty of Science, Technology and Environment.

Additionally, students wishing to progress past Bachelor level can include a postgraduate mathematics education paper as part of a Postgraduate Diploma in Education, and this Diploma qualifies a student to focus on a mathematics education topic for their MEd research project which may also focus on some aspect of mathematics education.

Other institutions

While there are other tertiary institutions in the region (see Table 5), most teach mathematics for specific vocational purposes, but all those involved in pre-service teacher education teach some mathematics education papers as part of their teacher preparation courses. However, most of these institutions are not involved in teacher education for senior high school

Table 5. Providers of Mathematics &/or Mathematics Education &/or Teacher Education.

Fiji	USP, Fiji National University, University of Fiji, Corpus Christi Teachers College
Kiribati	USP, Kiribati Teachers College
Samoa	USP, National University of Samoa
Solomon Island	USP, Solomon Islands National University, Solomon Islands College of Higher Education
Vanuatu	USP, Vanuatu Institute of Teacher Education

teachers. Some of these institutions were formed by the amalgamation of smaller institutions (and this amalgamation process is likely to continue). Currently USP is the foremost provider of tertiary academic education for the twelve countries in the region, and the only one to offer mathematics education as part of a Masters level qualification.

Concluding Thoughts

When writing about teachers and education, Nabobo-Baba (2006) acknowledged the different experiences that many of today's teachers have had since their childhood, and that change is inevitable and is global and local, as well as being both oppositional and complementary. Within this complex change process, Nabobo-Baba, along with other educators, has identified both the curriculum and the current teaching practices as needing to be reconceptualised to ensure education continues to be relevant to Pacific Island contexts and to the students' needs. Such a reconceptualization would involve: developing responsive pedagogies, listening to Pacific voices, finding alternatives to donor-driven reform (with donor-influenced goals), incorporating values education and local knowledge and wisdom, improving pre-service teacher education and professional development, acknowledging the influence of new technology and the media, designing culturally appropriate and future-oriented curriculum documents and resources, ensuring that indigenous pedagogies (rather than western ones) are used in teacher education (and in schools) and that these pedagogies are based on the values and ideologies of the indigenous culture of origin.

All these aspects of change relate to all subjects including mathematics and mathematics education; and to all levels of education (primary, highschool, and tertiary). But being practical, this agenda is too big, one cannot change everything all at once, and each island nation needs to decide on its priorities.

In summary, western mathematics educators believe that they have contributed much to international education, but they must also accept that they have ignored local cultural practices and colonized the minds of many people from different cultures in terms of mathematics and education. The alternative could have been that westerners learnt from the different local ways of being, knowing and learning—perhaps it is not too late.

Education systems typically change to reflect the values of the society. However, in the Pacific region the values of the current Island Nation leaders are different from those of the colonial leaders of 40 or so years ago, yet the education system continues to reflect many of the values of the colonists (which reflects the colonists' success in terms of colonizing minds). Having become independent nations, it will be interesting to see if education develops independently or continues to reflect western influences.

Recently, the historian Harari (2016, pp. 235–237), gave three 'formulae' for knowledge[1]. He wrote that the formula:

- in medieval Europe was
  Knowledge = Scriptures x Logic,
- during the scientific revolution was
  Knowledge = Empirical Data x Mathematics,
- and now with humanism is
  Knowledge = Experiences x Sensitivity.

It is interesting to consider the relevance of these to Pacific Island nations.
- Pre-colonial   education   seemed   based   on   humanism (experiences x sensitivity),
- Missionary education was certainly based on scriptures x logic (mainly arithmetic),

---

[1] Harari emphasised the need in these formulae for multiplication rather than addition because, if either factor was zero then the knowledge gained would be zero.

- Colonial education was empirical data (or information) x mathematics (with mathematics implying logic including the logic of associationism and behaviourism), and now
- Post-colonialism suggesting a return to humanism, may imply less mathematics.

In spite of colonialism, Pacific islanders' natural ways of knowing continue to be based on their experiences (formal and informal) and their sensitivity, which leads to their knowledge construction. It seems that Pacific education has travelled full circle. Whether this will lead to a de-emphasis of mathematics, or an integration of mathematics with other subjects, remains an open question.

References

Bakalevu, S., Tekaira, N., Finau, V., & Kupferman, D. (2007). Local knowledge and wisdom in Pacific teacher education. In: Puamau, P. (Ed.), *Pacific voices; teacher education on the move*. Suva, Fiji: PRIDE Project, Institute of Education, University of the South Pacific. pp. 69–83.

D'Ambrosio, U. (2001). What is ethnomathematics, and how can it help children in schools. *Teaching Children Mathematics (NCTM)*, 7(6), 308–310.

Fasi, 'U. (2005). The role of assessment in educational planning. In: Puamau, P., & Teasdale, G. R. (Eds.), *Educational planning in the Pacific; principles and guidelines*. Suva, Fiji: PRIDE Project, Institute of Education, University of the South Pacific. pp. 99–109.

Harari, Y. N. (2016) *Homo Deus; a brief history of tomorrow*, London, UK: Harvill Secker. (First published in 2015 in Hebrew as: *The history of tomorrow*, Israel: Kinneret Zmora-Bitan Dvir.)

Lewis, D. (1994). *We the Navigators: the Ancient art of Landfinding in the Pacific*, Honolulu, HI: University of Hawaii Press Inc.

Nabobo-Baba, U. (2006). Teacher education for new times, reconceptualising pedagogy and learning in the Pacific. *Directions: Journal of Educational Studies, 28*(1 & 2), 63–91.

Prasad, R (2004) *Tears in Paradise: Sufferings and Struggles of Indians in Fiji 1879-2004*. Auckland, New Zealand: Glade Publications.

Thaman, K. H. (2001). Towards cultural inclusive teacher education with specific reference to Oceania. *International Education Journal 2*(5), WCCES Commission 6 Special 2001 Congress Issue. http://www.flinders.eduau/education/iej

Thaman, K. H. (2009). Towards cultural democracy in teaching and learning with specific references to Pacific Island Nations (PINs). *International Journal for the Scholarship of Teaching and Learning, 3*(2), Article 6. http://doi.org/10.20429/ijsotl.2009.030206

Thaman, K. H. (2013). Quality teachers for indigenous students: an imperative for the twenty-first century. *The International Education Journal: Comparative Perspectives, 12*(1), 98–118.

United Nations. (1994). Declaration of Barbados. (Annex 1 of the Report of the "Global Conference on the Sustainable Development of Small Island Developing States", (held in Bridgetown, Barbados, 26 April–6 May, 1994). New York, NY: United Nations.

About the Authors

Andy Begg is an Associate Professor in the School of Education at Auckland University of Technology, New Zealand. He has taught high school mathematics, co-authored mathematics textbooks, and been a Government mathematics curriculum officer. Next, he taught, researched, and supervised postgraduate mathematics education students at the University of Waikato for ten years; continued such work for three years overseas; and then moved to his current position—working with and supervising post-graduate education students.

Salanieta Bakalevu is the Associate Dean of Learning and Teaching in the Faculty of Arts, Law and Education; and a Senior Lecturer in the School of Education at the University of the South Pacific in Suva, Fiji. Before joining the university faculty, she graduated from the university, taught mathematics in Suva to high school age students, and later completed her postgraduate studies at the University of Waikato in New Zealand.

Robin Havea is a Senior Lecturer in Mathematics in the School of Computing, Information and Mathematical Sciences, at the University of the South Pacific in Suva, Fiji where he teaches mathematics. He was born, attended school, and completed his first degree at 'Atenisi Institute in the Kingdom of Tonga. He moved to New Zealand for his postgraduate studies in pure mathematics, firstly at the University of Waikato and later at the University of Canterbury.

SOUTH PACIFIC: Appendix

1. Numbers

	1	2	3	4	5	6	7	8	9	10
Polynesia										
Cook Is	tai	rua	toru	ā	rima	ono	itu	varu	iva	tai ngaʻuru
Niue	taha	ua	tolu	fa	lima	ono	fitu	valu	hiva	hogofulu
Samoa	tasi	lua	tou	fa	lima	ono	fitu	valu	iva	sefulu
Tokelau	tahi	lua	tolu	fā	lima	ono	fitu	valu	iva	hefulu
Tonga	taha	ua	tolu	fā	nima	ono	fitu	valu	hiva	hongofulu
Tuvalu	tasi	lua	tolu	faa	lima	ono	fitu	valu	iva	sefulu
& NZ Māori	tahi	rua	toru	whā	rima	ono	whitu	waru	iwa	tekau
Hawaiʻi	kahi	lua	kolu	hā	lima	ono	hiku	walu	iwa	umi
Tahiti[1]	tahi	piti	toru	maha	pae	ono	hitu	vaʻu	iva	ʻahuru
Melanesia										
Fiji	dua	rua	tolu	vā	lima	ono	vitu	walu	ciwa	tini
Solomon Is	(Solomon Islands use many languages; Pidgin and English are the two most common.)									
Vanuatu	wan	tu	tri	fo	faef	sikis	seven	eit	naen	ten
Micronesia										
Kiribati	teuana	uoua	teniua	aua	nimaua	onoua	itiua	waniua	ruaiiua	tebuina
Marshall Is	juon	ruo	jilu	emen	lalim	jiljino	jiljilimjuon	rualitok	ruatimjuon	jiljino
Nauru	aiquen	aro	aiju	aeoq	aijimo	año	aelu	aoju	ado	ata

[1]Tahiti is the main island in the Society Islands which, along with the Austral Islands, the Tuamotu Islands, the Marquesas Islands, and the Gambier Islands, forms French Polynesia. The indigenous languages on these islands all vary somewhat.

2. Alphabets[1]

Polynesia

Cook Is	a	e	ng	i	k	m	n	o	p	r	t	u	v		
Niue	a	e	i	o	u	f	g	h	k	l	m	n	p	s	t	v
Samoa	a	e	i	o	u	f	g	l	m	n	p	s	t	v	h	k	r
Tokelau	a	e	i	o	u	f	k	g	l	m	n	p	h	t	v
Tonga	a	e	f	h	i	k	l	m	n	ng	o	p	s	t	u	v
Tuvalu	a	e	i	o	u	f	g	h	k	l	m	n	p	s	t	v

& NZ Māori	a	e	h	i	k	m	n	ng	o	p	r	t	u	w	wh
Hawai'i	a	e	i	o	u	ā	ē	ī	ō	ū	h	k	l	m	n	p	w
Tahiti	a	e	f	h	i	m	n	o	p	r	t	u	v

Melanesia

Fiji (pronounced as in brackets)

a(ar) b(mb) c(th) d(nd) dr(ndr) e(ay) f g(ng) h i(ee) j k kw l m
n o((g)o) p q(ng) qw r s t u((z)oo) v w y z

Solomon Is   English alphabet (a to z), English is the official language.

Vanuatu[2]

a ae b e f h I j k l m
n ng o p r s t u v w y

Micronesia

Kiribati

a b e i k m n ng o r t u w

Marshall Is

a ā e i o ọ ō u ū b d dw j k
l ḷ ḷw m m̧ n ṇ ṇw ñ ñw p r t w

Nauru

a ā e i o ō u ū b bw c d di f g gw h j k
kw n ng l m n ñ p qu r s t ti ts v w x y z

---

[1]Note that x, y, and z, the usual symbols used for variables in equations, are not included in all Pacific Island alphabets.

[2]Vanuatu has 3 official languages, English, French and Bislama, and 100⁺ local languages. This is the Bislama alphabet.

Chapter II

AOTEAROA/NEW ZEALAND:
Mathematics Education in
Aotearoa/New Zealand

Andy Begg
Auckland University of Technology, Auckland, New Zealand

Jane McChesney
University of Canterbury, Christchurch, New Zealand

Jyoti Jhagroo
Auckland University of Technology, Auckland, New Zealand

Introduction

New Zealand (NZ), a small country in the South Pacific, was first settled by Māori people who came from their Polynesian homeland about 800 years ago and named it Aotearoa. They came in large canoes and demonstrated navigational skills which suggest mathematical thinking, but their techniques were different from those developed by European navigators centuries later (Barton, 2008; Lewis, 1994). Before Europeans arrived, Māori elders had organized education for selected children in *whare wānanga* (houses of learning) where they learnt songs, chants, tribal history, spiritual knowledge, and knowledge of medicinal plants. However, for the majority of children, learning was informal and experiential. Mathematics was not taught or used, but evidence of mathematical thinking existed in counting, symmetry, measurement, patterns, buildings, and games.

The first European to reach NZ was Abel Tasman from Holland in 1642; he was followed by James Cook in 1769 who 'claimed' the country for Britain. Next, whalers and sealers came from numerous countries, and in the early 1800s missionaries and settlers arrived from Britain and Ireland. During this period a small group of Māori people travelled with Europeans to Sydney and visited a school; on their return they asked the missionaries to set up a school, and the first school was established in 1816 in the Bay of Islands (Jones & Jenkins, 2016).

In 1840 many Māori chiefs signed a treaty with the British crown, and a more formal association with Britain began. The Māori people quickly learnt the English language and how to trade and calculate using money.

By 1881 the European population was 500 000 and the colonization of land and minds continued. Before World War 2 most migrants came from the British Isles with free or assisted passage; a smaller number were from France, Germany, Scandinavia, China, and Dalmatia. After the war refugees came from Europe, and in the 1950s and 1960s many arrived from England, Scotland and Holland. Additionally, many Polynesians migrated from Samoa, Tonga, the Cook Islands, and other Pacific Islands. In the 1970s and 1980s government immigration policies changed, resulting in a significant number of immigrants from Asia and Africa. NZ's population in 2016 was 4.6 million and with diverse immigration and inter-marriage, many people now identify with more than one racial group; however, the dominant influence on education continues to be from Britain.

School Organization

In the 1800s New Zealand had primary schools with eight class levels for children from age 5 to 13, and high schools with up to five class levels for students of age 13 to 18. More recently the range of types of school has increased, and these are summarized in Figure 1. While most New Zealand schools are state-owned, a number of other schools also exist and these rely heavily on funding from the state.

Types of Schools in New Zealand

Age	School Yr	Class Name 1	School types
0–4	optional	pre-school	childcare, playcentre, and kindergartens
5	1	Junior 1	Contributing Primary School / Full Primary School / Composite or Area School (Yr 1–13) and Correspondence School (Yr 1–13)
6	2	Junior 2	
7	3	Standard 1	
8	4	Standard 2	
9	5	Standard 3	
10	6	Standard 4	
11	7	Form 1/Std 5	Intermediate School / Yr 7–13 Secondary School / Middle School (Yr 7–10)
12	8	Form 2/Std 6	
13	9	Form 3	Secondary (High) School
14	10	Form 4	
15	11	Form 5	Senior High School
16	12	Form 6	
17	13	Form 7	

Figure 1. Types of school in New Zealand

Mathematics in Schools

It is difficult to source the early history of mathematics education in NZ, however, the work of Simon (1994) provides an invaluable summary of early schooling.

1800s to 1920s

The first schools, established and managed by missionaries, churches and individuals, were based on the prior experiences of the missionaries and settlers from Britain and Ireland. The missionaries learnt the Māori language, and used it in the textbooks they produced and in teaching. They attempted to persuade Māori to give up their customs, habits, and values; and assimilate into the European culture, and convert to Christianity.

After 1852 the country was divided into six provinces with each having responsibility for their own schools; however, by the 1870s differences existed between schools across the provinces and people were thinking of the country as one nation. In 1877 primary schooling, providing the initial six or seven years of schooling, was made compulsory for all European children aged 7 to 13 (though often children started at age 5), and in 1894 this was extended to include all Māori children. In 1878 the first national school syllabus was written and it included arithmetic. For example, in the fourth year of schooling the work prescribed was:

> Numeration and notation of not more than six figures; addition of not more than six lines, with six figures in a line; short multiplication, and multiplication by factors not greater than twelve; subtraction; division by numbers not exceeding 12, by the method of long division, and by the method of short division; multiplication tables to 12 times 12; relative values and chief aliquot parts of the ton, hundred-weight, quarter, stone, and pound; relative lengths of the mile, furlong, chain and rod. Mental arithmetic adapted to this stage of progress. (Department of Education, 1900)

In 1867 the government began to set up 'Native Schools' for Māori in rural areas. This parallel but segregated system began to be phased out 60 years later, though a small number continued until 1969. Throughout this period many Māori families moved to urban areas where their children attended ordinary state schools and were assimilated within European culture. For most Māori children this English-styled schooling reinforced their subordinate status in their own land.

In 1903 the government provided free places in high schools for children (ages 13 to 17) who gained a Certificate of Proficiency at primary school. This led to the establishment of manual, technical, commercial, and academic pathways in high schools which 'filtered' professional and working-class children. The mathematics in the non-academic pathways was termed 'core' mathematics and was little more than arithmetic. In that period girls were not encouraged to learn academic mathematics, and this attitude persisted until 1943.

In 1922 the Correspondence School was established to provide distance education for primary school children living in remote areas and in 1928 this was extended to include high school education. Also, in 1922, the first 'intermediate' schools were established in cities in an attempt to improve the transition from primary school to high school; these were for year 7 and 8 students (previously, the last two years of primary schooling) and sometimes years 9 and 10 (the first two years of high school), that is, for students of age 11 and 12 (and sometimes 13 and 14).

1930s to 1950s

In the 1930s most students had left school at age 13, but this changed seven years later when the high school entry examination was abolished, though 'entry' tests were used to sort and segregate students into academic and non-academic courses.

Clarence Beeby (1902–1998) joined the NZ Government's Department of Education in 1939 and in 1940 became its Director. His thinking and mission was summarized by what he wrote in 1939 for the Minister of Education,

> *... every person, whatever his level of academic ability, whether he be rich or poor, whether he live in town or country, has a right,*

as a citizen, to a free education of the kind for which he is best fitted, and to the fullest extent of his powers.

With this perspective, he changed NZ's education. Under his directorship, a committee was set up to report on the Post-Primary School Curriculum, and their report (the Thomas Report) in 1943 led to a curriculum being developed for high schools. The report introduced a school leaving age of 15 and two new examination-based qualifications in years 3 and 4 of high school (school certificate and university entrance). Mathematics, together with English, science, social studies, and physical education, formed the common core curriculum for all students in their first two years of high school.

During the 50s arithmetic textbooks were provided by the Department of Education for primary schools. Most primary school teachers were not mathematics specialists, so these textbooks were closely followed; the focus was arithmetic though simple statistics and graph concepts were included. At the same time high schools were free to purchase the textbooks of their choice. In the 1950s there was also a huge increase in demand for schooling caused by the high birth-rate after the war, and by students staying longer at high school. This resulted in many more schools being built, in particular, large high schools.

1960s, 1970s and 1980s

Three changes in the sixties and seventies required new textbooks: decimalization of currency in 1967, metrication from 1969 to 1976, and the move from traditional to modern mathematics. These involved new curriculum documents (Department of Education, 1969; 1972) and contributed to an increase in locally authored mathematics textbooks.

The 'new-maths' movement began with a small number of high school mathematics teachers trialling ideas involving sets, Boolean algebra, vectors, matrices, and group theory. Initially, they drew on ideas from the USA's School Mathematics Study Group; and later on, ideas and approaches from England and Scotland that included transformation geometry. At the same time some teachers became interested in teaching more statistics. These initiatives were all encouraged by the Department of Education and involved many week-long courses involving mathematics teachers, inspectors, advisors, lecturers, and curriculum officers. With approval, schools could voluntarily opt for modern rather than traditional mathematics, though a few years later schools were told they must change to the modern curriculum.

By the mid-1980s modern mathematics was accepted by most teachers, and new curriculum documents were written for the first ten years of schooling. These included sets, relations, symmetry, transformation geometry, trigonometry, statistics, probability, and traditional arithmetic and algebra (Department of Education, 1985, 1987). This curriculum content was extended for the final three years of schooling with changes in the examination syllabi for years 11, 12, and 13. These initiatives were supported by teachers guides for high schools and new primary school resources that included Cuisenaire rods, Dienes blocks, and new textbooks (student and teacher versions).

The fourth significant change involved technology, first calculators, and later, computers. Simple four-function calculators came first, followed by scientific calculators, later in the 1980s graphics calculators, and after those, CAS (Computer Algebra System) calculators were permitted in class and in examinations. This meant that there was no longer a need to teach students how to draw graphs or calculate standard deviations as these could be generated automatically from the input of datasets. Unfortunately, this created an environment in which some students had advantages over others in terms of assessment because they could afford the more sophisticated calculators.

The fifth was the growing interest in computer programming, the sixth involved more emphasis on statistics, and both these changes were also occurring in universities. The increased emphasis on statistics and the formal introduction of computing first occurred at the senior level in applied and additional mathematics courses which had previously focussed on mechanics. Over time statistics became a part of school mathematics at all levels, mechanics (which was also taught in physics) was dropped from mathematics, and computing became a separate school subject. This led to two new courses in the final year of schooling, 'Mathematics with Statistics', and 'Mathematics with Calculus'. The calculus paper was considered the more difficult, but was the pre-requisite for some university courses.

The seventh initiative, begun in the 1980s with the Department of Education's support, was the development of resources so that mathematics could be taught using the Māori language. This involved working with the Māori Language Commission to develop a mathematics vocabulary, with one of the first words being the word for mathematics itself. The word chosen was pāngarau, with 'pānga' meaning 'relationships', and 'rau' meaning 'many' (literally, 'hundred'). This work made all involved conscious of the fact that in pre-colonial times in NZ (and the Pacific Islands) there had been no such thing as mathematics beyond simple counting. For Māori, shapes, patterns, and symmetry were not geometry, they were part of design, art, and craft; and belonged to craftspeople, not to mathematicians; and, to consider them as mathematical was to take away the 'mana' (prestige) of the craftspeople. At the same time, many involved in the Māori-mathematics language above development project were interested in ethnomathematics. Respecting the Māori perspective, mathematics educators have not used traditional artifacts as examples in mathematics, though there has been an acknowledgement of Māori ways of learning by doing, listening, and observing, rather than the English/European emphasis on reading and writing.

From 1960 until 1989, the developments were supported by the Curriculum Division of the Department of Education. Curriculum officers organized week long live-in workshops with teachers seconded from schools to work on new curriculum documents and develop resources for teachers. At the same time, New Zealanders were looking overseas for ideas, and inspiration—overseas textbooks had been commonly used, but international visitors had not been common. As part of 'internationalization', a major catalyst occurred in 1984, a number of New Zealanders attended ICME when it was held in Australia; and this was followed by regular contact with the Mathematics Education Research Group of Australia and with other international mathematics education groups.

In 1989 the Department of Education was replaced by a Ministry of Education. This change was based on the assumption that curriculum and professional development would be contracted out to teachers and specialists working on short-term projects. It resulted in not having a permanent group of subject experts at national level to give advice to government or coordinate curriculum and resource development. At the same time that the Ministry was established, two other government agencies were set up, the Education Review Office (ERO) that took over the role of school inspectors, and the New Zealand Qualification Authority (NZQA) that took over the coordination of all school-leaver qualifications including school certificate (year 11), university entrance (year 12), and university bursary and scholarship (year 13).

1990s

One change that began in a small way in 1990 involved postgraduate mathematics education qualifications being offered for practising teachers. Since this time many people have graduated with masters degrees and doctorates that focus on mathematics education.

Another change involved the restructured Ministry of Education contracting out the production of a new mathematics curriculum for the first ten years of schooling. This combined the primary school and high school mathematics curriculum for the first time, though the final three high-school years were covered by the syllabi for external examinations for qualifications. This curriculum, published in 1992, included 'General Aims of Mathematics Education.' 'Mathematical Processes' (problem solving, developing logic and reasoning, and communicating mathematical ideas) that complemented the traditional content knowledge, and five learning areas (number, measurement, geometry, algebra and statistics), that were to be taught to all students from age 5 onwards. Many teachers were surprised by the inclusion of statistics, geometry and algebra being included from the start of schooling for 5-years-old learners. Many of the ideas in the curriculum were stimulated by contact with colleagues from Australia, England, and the Netherlands. Sample assessment activities were included that reinforced the importance of assessment but these activities emphasized content and ignored the processes, consequently teachers changed the content of their programmes but ignored the aims and processes which better summarized the spirit of this curriculum.

Two years later the first official curriculum for mathematics in the Māori language, *Pāngarau* (Ministry of Education, 1994), was published. This was a new curriculum rather than a translation of the 1992 English-medium Mathematics Curriculum, although the mathematics content was very similar.

In between these two publications, and before other subject curriculum documents were developed, *The New Zealand Curriculum Framework* (Ministry of Education, 1993) was distributed. This drew together seven 'essential learning areas' (languages, mathematics, science, technology, social sciences, the arts, and health and physical well-being), and listed eight 'essential skills' that were common to all subjects (communication skills, numeracy skills, information skills, problem-solving skills, self-management and competitive skills, social and co-operative skills, physical skills, and work and study skills). However, these skills were not assessed so were often ignored.

Another initiative that had begun in the Department of Education in the mid-eighties and continued with the Ministry was the *Beginning School Mathematics* project for the first three years of primary school (ages 5 to 7). Resources for teachers were produced and distributed in the 1990s with information about new content—statistics and the mathematical processes (problem solving, logic and reasoning, and communicating mathematical ideas). These resources included teacher-led activities and independent tasks, that focused on mathematical activities, games, and puzzles; and resulted in all schools having a consistent set of mathematics equipment. Many teachers used and adapted these resources to teach small groups of eight to ten students while others in the class worked on independent activities. For younger learners, these resources helped shift the focus from a textbook or blackboard approach to a hands-on and structured approach that encouraged children to explore concepts in informal ways.

2000 to 2016

Because of concerns arising from the second TIMSS study, in 2000 the Ministry of Education prioritized numeracy and literacy for schools. They implemented the Australian *Count Me In Too* programme that was based on Steffe and Cobb's work and included schedules for assessing early number learning, using interview-based tasks. After trialing in 81 schools this was expanded into a Numeracy Development Project (NDP) and rolled out across different levels of the primary school. This involved numeracy facilitators working with teachers from groups of schools who had opted into the associated professional development. The NDP included a "number framework" that set out two domains of number learning, namely strategy and knowledge, and identified eight 'stages' of progression of number learning; and resources were produced to support the programme. Aspects that were controversial included: splitting of mathematical learning into strategy and knowledge, direct teaching of specific mental calculation strategies, and stages of learning (especially the early counting stages which became a barrier for many students). One outcome of the project was grouping of students by ability either within their class or between classes, (a reintroduction of a form of streaming). The programme advocated a specific lesson sequence based on group teaching and independent learning that contrasted with the research evidence presented by a 'best-evidence' synthesis of research related to 'effective pedagogy' (Anthony & Walshaw, 2007) which focused more on meeting the range of student needs. The NDP Number Framework became a proxy curriculum for number, and continues to influence primary school mathematics, with resources now available from the Ministry website.

In 2007 the Ministry of Education distributed a radically new curriculum. It covered all subjects (English, the arts, health and physical education, languages, mathematics and statistics, science, social sciences, and technology) rather than having separate documents for each. It spanned the continuum of classes from years 1 to 13, and included Values, Principles, and Key Competencies. The key competencies were: *thinking; using language, symbols, and text; managing self, relating to others; and participating and contributing.* However, little was done by teachers to address these aims within mathematics classrooms. The focus remained on teaching content knowledge for assessment. The mathematics part of this curriculum began from year one with three focus areas in mathematics: number and algebra, geometry and measurement and statistics for all students from age 5 onwards. Perhaps the biggest difference between this mathematics curriculum and overseas ones was the emphasis given to statistical investigations (that began at age 5) and built up to include using the statistics enquiry cycle which involved: regression, time series, exploratory data analysis; and inferences involving confidence intervals, central limit theorem, and resampling, as well as margins of error, probability, and the Poisson, binomial and normal distributions. In addition, the mathematical processes (problem solving, logic and reasoning, and communicating mathematical ideas) continued to be emphasized. This was followed by *Te Marautanga O Aotearoa* (Ministry of Education, 2008), the curriculum for Māori-medium schools.

A change in government saw the introduction of national assessment in primary schools with 'national standards' being introduced for reading, writing and mathematics. A year later *Mathematics Standards for years 1 to 8* (Ministry of Education, 2009) was published—it was intended as a teachers' guide, but it emphasized assessment rather than the key competencies, and much of it was modeled on the NDP Number Framework.

A new three-level National Certificate of Educational Achievement (at Levels 1, 2, and 3) was introduced for years 11, 12 and 13 school leavers between 2002 and 2004, though the existing scholarship examination continued for high-achievers in year 13. This 3-level national certificate replaced the School Certificate, University Entrance, and University Bursary examinations and it increased the emphasis on assessment and reflected an attitude of academic elitism which ran counter to the egalitarian attitudes of many. Content was categorized into internally or externally assessed 'standards', and teachers were expected to use competency-based assessment processes. This certificate has led to a greater atomization of mathematical programmes; and to content knowledge being given more emphasis at the expense of the mathematical processes (from the 1992 curriculum), the essential skills (from the 1993 curriculum framework), and the key competencies (of the 2007 curriculum). One reaction to this qualification has been a move by some schools to the 'International Baccalaureate' or the 'Cambridge International Examinations.'

Calculators and computers have been used in schools for over 30 years, but the influence of technology became more evident after 2010 with a BYOD (Bring Your Own Device) mandate coming from many schools. Now, computer and e-based resources are increasingly replacing textbooks and exercise books; and enabling more communication between teachers, students, and parents. This has resulted in the defection of textbook publishers from our country which implies a shift from local to international educational resources and this may be a future concern. While e-learning brings new opportunities, it also raises concerns, but over time it is expected to be fully accepted.

Pre-school education

Most NZ children start school at age 5, having previously attended non-compulsory child-care centres, play-centres, or kindergartens. Traditionally these institutions were not concerned with school subjects, but this began to change in the early 1990s when the Ministry of Education funded research to investigate educational activity in early childhood contexts. The research focused on familiar contexts, meaningful purposes, and shared mathematical activities; and fed into the development of our first early childhood curriculum, 'Te Whāriki' which had five strands— well-being, belonging, contributing, communicating and exploring (Ministry of Education, 1996). Within these strands, the curriculum explanations were not intended as a teaching programme, but rather, as a guide for early childhood teachers as they provided learning opportunities.

To support this curriculum, a resource for teachers called *Making things count* (Ministry of Education, 1999) was produced. It provided examples, ideas and resources for identifying and enhancing mathematical experiences and learning in early childhood centres. Other resources were developed including a mathematics framework for early childhood education based on the metaphor of *te kākano* (the seed); this framework acknowledged that young children bring to and develop considerable knowledge in their early childhood education; and was first published in a resource for teachers that used three lenses—assessment, curriculum, and mathematics. Another resource, *Te Aho Tukutuku—Early mathematics* (2012) was also developed by a Ministry of Education working group to support early childhood mathematics, the mathematics involved was in six strands—summarized by the Education Review Office (2016, p. 8) as:

- Pattern - the process of exploring, making and using patterns
- Measuring - answering the question, "How big is it?"
- Sorting - separating objects into groups with similar characteristics
- Locating - exploring space or finding or 'locating' something, such as a place (location), or an item in space
- Counting and grouping - the process for working out the answer to a question about "How many?" Grouping involves putting things together.
- Shape - naming shapes and identifying the unique specific properties or features of shapes.

Today NZ faces a number of challenges including:

- mathematics teachers are becoming an aging profession, who will replace them?
- many Māori and Pasefika students are disenfranchised, what needs to be done?
- students at all levels continue to be assessed by examinations, is there a better way?

University Mathematics and Mathematics Education

There are currently eight universities in NZ. The first, the University of Otago, was established in 1869. However, in 1874 the University of New Zealand was formed as our sole degree-granting university and it incorporated the University of Otago together with the other university colleges and two agricultural colleges in the following years.

Since the dissolution of the University of New Zealand in 1961 these colleges have become independent degree-granting universities, two new universities have been established, and the previously autonomous teachers' colleges have all been incorporated into our current eight universities.

Mathematics

The first mathematician to be appointed as professor in NZ was John Shand (1834–1914). He was appointed as professor of mathematics and natural philosophy (physics) at the University of Otago and arrived in 1870, a year before the university officially opened in 1871. He was an outstanding teacher, and served the university well until 1913, initially concentrating on teaching, but later as an administrator.

In the next few years, other university colleges were set up and in the first hundred years our university mathematics courses closely followed British traditions. Five notable mathematicians from these early years were Maclaurin, Bell, Forder, Aitken, and Bullen.

- Richard Maclaurin (1870–1920) came from Scotland at age 4, completed his masters degree at Auckland in 1890, then studied at Cambridge and Strasbourg. In 1899 he became chair of mathematics at the newly created University College in Wellington. Eight years later he accepted an offer from Columbia University but in 1909 he took up the position of President of the Massachusetts Institute of Technology where he worked until his death.
- Robert Bell (1876–1963), known particularly for his textbook on 3-dimensional coordinate geometry, was appointed professor of mathematics at Otago in 1919 and taught there until he retired in 1948.
- Henry Forder (1889–1981) born in England, studied at the University of Cambridge and taught in high schools in England. He came to Auckland and was Professor from 1933 to 1955. In addition to leading the department, he was well known outside the university for his school geometry textbooks.
- Alexander Aitken (1895–1967) was born in Dunedin, NZ, and studied at the University of Otago before going to the University of Edinburgh where he gained his PhD. He is best known for his work

in algebra, numerical analysis, and statistics. During WWII he worked in the team at Bletchley decrypting the ENIGMA code.
- Keith Bullen (1906–1976) was born in Auckland, NZ, studied at the University of Auckland, went on to Cambridge for his doctorate; taught at universities in Auckland, Cambridge, Hull, Melbourne, and Sydney, and specialized in applied mathematics, in particular, seismology.

Mathematics in our universities has always been evolving and this is influenced by trends in other English-speaking universities. While the content of many university papers has changed over the years, the teaching and assessment has not changed markedly, except that technology is now taken for granted. One significant change has been the evolution of geometry. Initially, and until the mid-50s, it was taught at undergraduate level using a traditional Euclidean approach. By the 60s the approach was with coordinates and occasionally vectors, and topology was introduced. Since the 80s geometry has not been emphasized at the undergraduate level, and is rarely taught now in our universities. Another change was the shift of applied mathematics (mechanics) from mathematics departments to physics and engineering. During this period both statistics and computing which began in mathematics departments in the early sixties, had grown in popularity and split from mathematics.

A report (Hunter, 1998), commissioned and published by the Ministry of Research, Science and Technology emphasized (among many other things):
- the important role of mathematics in underpinning almost all activities in modern society
- the need for a high level of quantitative literacy within a knowledge-based society
- the recognition of the proper uses of computer software, and
- the increased professionalism of mathematics and statistics.

Additionally, this report was concerned about the continuing shortage of well-qualified mathematics teachers in high schools, and the small pool of teachers with sufficient mathematical knowledge in primary schools.

Many tertiary mathematics papers (i.e., units of study) continue to be listed with traditional names such as: algebra, calculus, topology, statistics, and probability; while other papers are linked to other subjects and are taught outside of the mathematics departments. For example, mathematical and statistical topics are included in courses such as operations research, biotechnology, geophysics, health sciences, architecture, and population studies.

NZ's reputation in the international mathematics community has grown over the years; while some mathematicians have worked mainly in NZ, others have been based overseas; two of the most notable are:

- Roy Kerr (1934–) born in NZ, studied at Canterbury University College in Christchurch, completed his doctorate at Cambridge in 1959, worked for the US Air Force until 1962, then moved to the University of Texas at Austin. He returned to the University of Christchurch in 1971 and remained there until he retired in 1993. He is best known for Kerr geometry which provides an exact solution to Einstein's field equation of general relativity. In 2013 he was the first New Zealander to be awarded the Albert Einstein medal.

- Vaughan Jones (1952–) studied at the University of Auckland and the University of Geneva, and won the Fields Medal in 1990 (the only New Zealander to have done so to date). Since 2011 he has been a Distinguished Professor at Vanderbilt, Professor Emeritus at the University of California (Berkeley) and a Distinguished Alumni Professor at the University of Auckland.

The research interests of NZ mathematicians are diverse and illustrate how mathematics is both pure and applied, and that partitioning knowledge into subjects, and subjects (such as mathematics) into topics, creates artificial barriers within knowledge when it is holistic. Such 'partitioning' is particularly evident in the dominant western ways of thinking in education and in the way courses are offered.

Statistics

Statistical work—collecting, analyzing and presenting data—has occurred in NZ since the British colonists arrived in the mid-1800s and annual statistical reports have been published by the government since the 1850s. Within the universities, statistics was first taught as a service subject in a number of sciences and social sciences. In the late 1950s and early 1960s the mathematics departments began to offer statistics courses; statistics as a stand-alone subject flourished, and statistics departments are now as large or larger than the mathematics departments.

The interest and focus on statistics in the school curriculum since the late 1960s has given rise to a number of important initiatives in NZ statistics education. Statisticians from universities and industry contributed to curriculum development by advising on the new curricula, initiating research in statistics education, developing student and teacher resources, and providing professional learning opportunities for teachers. Two notable contributors to statistics and to statistics education have been:

- Geoffrey Jowett (1922–2015) arrived in NZ in 1964 and became Professor of Statistics at the University of Otago. He later moved to the Biometrics Section of the Invermay Agricultural Research Centre. Geoff contributed significantly to the introduction and establishment of statistics education at Otago University and in NZ high schools.
- David Vere-Jones (1936–) was born in London, studied in NZ at Wellington, completed his doctorate as a Rhodes Scholar at Oxford, and became Professor of Mathematics at Victoria University of Wellington in 1970. He led the Institute of Statistics and Operations Research, supported the introduction of statistics into the NZ high school curriculum. From 1987 to 1989 he chaired the Education committee of the International Statistical Institute.

Computing

In the early 1960s and 1970s computers were being developed and a number of university mathematicians became interested in how they worked, how they could be used to solve mathematical problems, and the development of computer programming. Many of the first university courses in computing were offered by mathematics lecturers, but as the interest in computing broadened, they came to be seen as more than a tool for mathematicians and computer studies became recognized not as part of mathematics, but as a discipline in its own right. Currently university computer science departments rival mathematics departments in size.

One NZ computer scientist, Tim Bell (1962–), designed *Computer Science Unplugged* as an accessible way of engaging children from as young as 5-years old with computer science, and this has an international following with resources published in a number of languages.

Mathematics education

While all eight universities teach mathematics, in the past only six offered mathematics education classes that were not part of pre-service teacher education. Pre-service education had occurred in separate state tertiary institutions, but these have now all merged with universities and currently seven of our eight universities offer teaching qualifications as well as general education courses for other degrees. (Additionally, there is also a small number of private teacher education providers.)

Mathematics education, as distinct from mathematics, has taken two main forms. Historically it was part of initial teacher-education; and this continues today. Additionally, since the 1980s it has been offered at undergraduate and postgraduate level and can be taken either before deciding to be a teacher or by teachers as part of their professional development. Such papers focus on aspects of education that relate to the learning and teaching of mathematics rather than on specific mathematical content. These papers are offered for initial degrees, higher degrees, and for professional development; and cover many topics including theoretical, political, and practical perspectives of teaching, learning and assessing mathematics across all educational sectors. There has been a long-standing commitment to researching mathematics education in multicultural classrooms, and this commitment has been reflected in a focus on ethnomathematics, and culturally responsive pedagogies.

Many people have been involved in mathematics education over the years. Two of the more notable contributors from universities have been from Gordon Knight (1934–) from Massey University and Derek Holton (1941–) from Otago University. Gordon worked with teachers throughout his career, served on numerous mathematics curriculum committees, and worked with Māori educators to identify mathematics related to cultural artifacts and traditional practices. Derek was known to many NZ teachers because of his support for mathematics education, his problem-solving resources, and his work for mathematics competitions and the Mathematics Olympiad.

Three others were Bevan Werry (1935–1987), Andy Begg (1940–) and Bill Barton (1948–). Bevan had been a high school teacher and a mathematics educator at Christchurch Teachers College before joining the Curriculum Development Division of the Department of Education and later becoming its Director; he is acknowledged by his colleagues for professionalizing mathematics education in New Zealand. Andy was a high school teacher, textbook author, publisher, curriculum officer, and the first mathematics educator appointed to teach and research mathematics education at the postgraduate level rather than for pre-service teacher education. He has worked at three NZ universities and for shorter periods at universities in Australia, Pakistan, and England. Bill was also a high school teacher, who contributed to the development of Māori mathematics education. His research interest was ethnomathematics. He became Professor of Mathematics Education at the University of Auckland, and was elected President of the International Commission on Mathematical Instruction from 2009 to 2011.

The interest and focus on statistics in the school curriculum over the last 40 years has given rise to a number of important initiatives in NZ statistics education with university and industry research statisticians contributing to curriculum development. Two statistics educators who contributed significantly are Maxine Pfannkuch (1949–) and Christopher Wild (1952–), both from Auckland University. They have been influential in the move to include more inferential statistics in the high school curriculum and for developing resources for schools. In addition, Megan Clark (1946–) and Sharleen Forbes (1948–) contributed by disseminating research for teachers that widened the focus on the social and cultural implications of statistics education.

Teacher education

There are two forms of teacher education, pre-service and in-service. In 1965 there were nine teacher training colleges (later called colleges of education) that provided pre-service teacher education; the first two having been established in the 1870s. By 1990 two of the nine had been closed and two had amalgamated. Over the next two decades all six colleges merged with their local universities, and the newest university also offered teacher education qualifications. These seven universities now provide initial teacher education for approximately 95% of high school teachers, 90% of primary school teachers, and 45% of early childhood educators, the balance being provided by small private teacher education providers.

The mathematics education courses within initial teacher education vary across the universities with the typical focus being the practical aspects of learning, teaching and assessing mathematics, and on the curriculum. Most teachers in primary schools (years 1 to 8, ages 5 to 12) teach all subjects and their initial qualifications usually involve a three-year degree programme that includes one or two mathematics education papers based on both mathematical and educational content. On the other hand, high school mathematics teachers have usually completed a mathematics and/or statistics degree, and their teaching qualification is generally a one-year course that focuses on teaching and learning.

Initial teacher education qualifications are currently in a state of flux with various models being piloted such as postgraduate entry into professional registration as a teacher.

In-service or continuing mathematics education takes numerous forms apart from participation in university study. Before 1989, the Department of Education had provided week-long residential 'refresher courses' during school vacation time that teachers could attend, and typically when a new curriculum was introduced, numerous courses were made available. In addition, regional mathematics associations, comprising mainly high school teachers, hosted regular meetings as precursors to change and as vehicles to support change. In the past the Department of Education had also organized working parties that involved a number of practising lead-teachers, these came together regularly for week-long working groups to work on curriculum documents and support material. These materials were published by the Department of Education and distributed to schools as part of the curriculum implementation process, and the teachers who had been involved often shared their expertise at teacher 'refresher' courses and other in-service meetings. There were also primary and high school mathematics advisors whose roles involved working in individual schools and running in-service courses. Advisors were crucial for supporting teachers in rural schools that were often one or two teacher schools and isolated from in-service opportunities in urban areas.

Since 1989 such support from the government has not been provided. The biennial conference of the NZ Association of Mathematics Teachers provides invaluable support, the regional mathematics associations host lectures and workshops for teachers, but schools are expected to take responsibility for development and to hire expertise from private organizations that offer consultancy services.

Mathematics Education Issues

The issues of concern to teachers of mathematics (and other subjects) include: gender, ethnicity, Māori education, Pasefika education, immigrant education, the ability range, professional development, teaching and learning, assessment, and technology. While some of these have been the subject of mathematics education lectures and the focus of research for a number of years, all continue to have the status of 'work-in-progress'. One major research project completed in 2007 was the *'Best Evidence Synthesis Mathematics/Pāngarau'* project by Glenda Anthony (1955–) and Margaret Walshaw (1950–); this was a major synthesis of research that was later published in summary form by the International Bureau of Education. Glenda has also worked with Roberta Hunter (1947–) in teacher development related to implementing 'powerful pedagogies', this involved orchestrating discussion and worthwhile mathematical tasks;

while Margaret is known internationally for her critiques of mathematics learning and teaching from post-modern perspectives.

Gender

Until 1950 high school mathematics was perceived as being for boys; many girls were expected to take subjects such as home economics and typing, and perhaps 'core' mathematics (basically, arithmetic). This situation was exacerbated by the shortage of high school teachers of mathematics, especially women. To remedy this situation in 1957 the Government introduced a three-year non-university course aimed to attract more women into the teaching of mathematics and science; but the course did not appeal to the desired number of participants and was later discontinued. In the 1960s NZ's women's liberation movement inspired many young women to seek employment within the professions, and this has involved many more women becoming mathematics teachers, heads of departments and school principals.

In the 1980s a chapter of the International Organization of Women in Mathematics Education (IOWME) was established in NZ. Helen Wily (1921–2009) was an early contributor and many women were involved in researching gender issues concerning mathematics education, and ensuring that gender was a focus in the analysis of the mathematics results of the early International Educational Achievement studies. Another focus on gender and mathematics (and science) began in the mid-1980's with the formation of EQUALS NZ—EQUALS being a way of thinking about equity, mathematics and learning. This group was influenced by the EQUALS work in the USA which included work by women involved as mathematics teachers, teacher educators, mathematicians, and statisticians. Three notable contributors to these issues in New Zealand were Jill Ellis (1936–), Barbara Miller-Reilly (1943–), and Maxine Pfannkuch (1949–).

Ethnicity

Since 2000 New Zealand's population mix has changed, and change is predicted to continue as shown in Table 1. The increasing percentages for

Asian, Māori and Pacific people suggest an increasing need to think about multicultural rather than mono-cultural or bicultural education. Such thinking needs to be with respect to subject-content, applications, teaching, learning, assessment, and the language of instruction. Some of the main cultural differences relate to:

- language – English or home language, technical language adequacy
- modes of teaching – telling, questioning, project-work, inquiry
- ways of working – individual, group, whole class
- grouping – gender mix, self-selected, ability, mixed-ability
- homework – prescribed, open ended, optional, none
- cultural attitudes – teacher status, asking and answering questions.

Māori education

The early missionary schools had taught using the Māori language but since 1903 this was discouraged, and the English language was promoted as being more appropriate for education. While Māori people valued education, many Māori students did not experience success in NZ's English-medium school in mathematics or in other subjects. This lack of success has been recognized since the 1960s. Some teachers and their communities tackled this issue by organizing Māori-language schools where the teaching of all subjects was in the Māori language. This ran counter to the nationwide policy that had existed from the early 1900s through to the 1970s of discouraging te reo (the Māori language) being spoken in schools; but it was supported by the increasing public demand for the acceptance of the right for Māori students to learn in their own language. As a result, by 1997 there were 675 kōhanga reo (pre-school

Table 1. NZ's population trends by ethnicity.

Year	2013	2025	2038
NZ Population	4.88m	5.72m	6.47m
European (& other)	68%	62%	56%
Māori	14%	15%	17%
Asian	11%	15%	18%
Pacific Islanders	7%	8%	9%

language nests) catering for 13 505 students; 54 kura kaupapa Māori (schools) and three wānanga (tertiary institutions) which provided Māori-medium education for 32 000 students.

For mathematics (and for other subjects) this had created a problem. No Māori vocabulary existed for most school mathematics. Two initiatives in 1991 helped alleviate this. The first was the publication of *Ngā kupu tikanga pāngarau*, a mathematics bilingual vocabulary list prepared by Bill Barton (in 1991) working with the Māori Language Commission. The second was a UNESCO-funded Conference (Mathematika Pasefika) hosted at the University of Waikato to foster the development of a mathematical vocabulary database (Begg [Ed.], 1991) in the languages of NZ Māori, Cook Islands, Fiji, Niue, Samoa, French Polynesia, Tokelau and Tonga.

With the emergence of Māori immersion schools and Māori-language mathematics classes in other schools, there was a considerable demand for teaching and learning resources; though now this demand seems to be catered for by resources from websites.

Additionally, cultural influences on education involve more than language. They include an emphasis on oral and visual rather than written communication, and a preference for group rather than individual learning; but these issues have been left for teachers to address.

Pasefika education

Currently NZ has a significant number of students from the Pacific Islands, however, most teaching is in English. This is due to tradition (schooling was influenced by colonists from England); to the reliance on English-language textbooks; to the lack of indigenous mathematical vocabularies; to the dominant position of English language in NZ; and to English being the only language of many teachers. When Pacific Island people come to NZ they assume that mathematics (and all other subjects) will be taught in English, however, more could be done to ensure that at least the pedagogy is culturally appropriate. Such a pedagogy would involve more emphasis on collaborative work rather than individual activity, and on oral and visual approaches rather than emphasizing the written word.

One mathematics educator who is particularly concerned with Pasefika children is Roberta Hunter. In her project the children are divided into communities of mathematical inquiry. They work together, discuss, and argue about mathematical concepts and problems related to culturally familiar objects. For example: the weight of a taro, the dimensions of a tapa cloth, and patterns on a piece of tivaevae cloth. The key being children investigating and learning from culturally familiar concepts, rather than teachers telling from textbooks.

Immigrant education

NZ has had immigrants arriving since the early 1800s. For the first 150 years the traditional British classroom practices that prevailed suited the migrant children more than the Māori students. Since the early 1950s more immigrants have arrived from the Pacific Islands, and since 2000, from Asia; and students from both these groups often assume that teaching is telling, that teachers should not be questioned, and that students should not discuss ideas in class; rather, that they should remember and be able to repeat what they have been told. Two assumptions underpin this belief— that the teacher is the expert, and that understanding evolves over time. This practice is reinforced by teachers who see their role as 'lecturing'; rather than 'teaching' or stimulating learning. This 'telling' approach runs counter to the current view that teaching should start where the learner is, and involve: asking, discussing, and discovering.

The reluctance of some students from different cultures to discuss is reinforced by their awareness that they do not fully understand, and that they may be ridiculed or 'lose face' if they give incorrect responses or make unjustified conjectures. Many teachers are becoming more aware of this and organize group work so that students respond for their group rather than as individuals as this partially overcomes this difficulty. Unfortunately, immigrant students are sometimes placed in low-ability classes or withdrawn from class because of their lack of proficiency with English, and while this may be well intended, it is viewed negatively by the students.

The government's change to immigration policies has also led to many immigrant teachers seeking jobs, however, the teaching practices that immigrant teachers commonly have are framed by the ideologies of teaching and learning from their home countries. Many see their work as knowledge-transmission, they expect to be respected as the source of knowledge and to be able to teach silent students that do not disrupt or question them, and many have difficulty shifting from their traditional teaching style. Although overseas-qualified teachers require teacher registration prior to being employed, and this involves having teaching qualifications and experience assessed and meeting appropriate language requirements, it does not ensure that the teachers have the ability to cope with the significantly different attitudes of many students in NZ.

Ability range

In the past, in many schools, students were 'streamed' into classes or 'grouped' within a class on the basis of ability. These practices continue in mathematics more than in other subjects, though often renamed as 'interchange' or 'cross grouping'. These practices were encouraged by the introduction of a 'core mathematics' option, and with alternative qualifications such as the 'local mathematics certificate'; but now, with assessment at all levels, teachers are expected to change their practice from whole-class teaching to a differentiated-teaching to accommodate diversity—but many find this difficult and continue to move students from one group or class to another for mathematics lessons.

Professional development

Until 1989 most professional development initiatives were provided by the government's Department of Education. These included the provision of teachers' guides, teacher refresher courses, study leave bursaries for teachers, and mathematics advisors in the regions. In addition, meetings and workshops were provided by the network of mathematical associations, the New Zealand Association of Mathematics Teachers, and the universities in the main city centres.

In 1989, when the Department of Education was replaced by a policy-focused Ministry, these opportunities were partially replaced by alternatives provided by commercial groups funded by government contracts. These alternatives were aligned with specific Government policies rather than with teachers' issues, and they were usually short-term while change typically takes a considerable time. However, special interest groups including the mathematics associations and the universities, continued to organize meetings, lectures, workshops and conferences that provide some professional development for teachers; and other resources and development activities are readily available on the web.

Teaching and learning

Traditionally the teaching of mathematics in schools was based on *direct instruction* or *'teaching as telling'*. However, in the last 100 or so years schools have been influenced by theories including associationism, behaviourism, discovery learning, and constructivism, and these continue to influence teaching. Associationism and behaviourism were often interpreted as requiring 'drill and practice' which is still evident in many classrooms, and the splitting of a topic into behavioural objectives to focus a lesson (and assessment) is also evident.

While many other theories abound, a more important emerging influence is not theory, but rather, architecture. Large open-plan flexible learning spaces (modern learning environments) are being provided in many new schools and in new buildings of existing schools. These spaces are shared by a number of classes and teachers, often teaching different subjects; and the architecture is intended to shift the emphasis from teachers-teaching to learners-learning, and the learning from being individually focussed to group focussed with groups sometimes spanning class levels. (Though such a focus does not fit easily with the government's continuing focus on high-stakes assessment.)

Regardless of theories or architecture, the teaching of mathematics, in particular in high schools and universities, has been dominated by symbolism and logic. While this has been the traditional approach to the subject it is not always the most appropriate way to build the confidence and understanding of all students when a topic is first introduced. Visual approaches (diagrams, arrow graphs, and other pictorial representations) seem rarely to be employed when new and unfamiliar topics are being introduced, yet visual proofs and intuitive thinking often help learners develop understanding more than line-by-line symbolic logic proofs. In addition, both Māori and Pacific cultures are oral and practical rather than written and theory-based, and this needs to be factored into education.

Assessment

Teachers informally assess students' work in diagnostic and formative ways to inform their teaching; and in summative ways with tests or examinations when formal reporting requires it. Such assessment is usually school-based (except for the final three years of high school when external assessment for various qualifications is the focus). More recently, with the numeracy project, more emphasis has been given to assessment at all levels, and teachers often over-assess. If students are being prepared for life-long learning then there is a need to shift from marks, grades, or competency-based labels that evaluate, to comments and questions that suggest ways forward, and from teacher-driven assessment to students' self-assessment which links with meta-cognitive thinking.

21st century technology

In the early 2000s teachers were issued with laptops by the Ministry of Education and the expectation was that the laptop would become an important tool for teaching and learning. More recently most students responded positively to the BYOD (bring your own device) call, though no funding was provided for families that could not afford such devices. In 2010, one commonly-used textbook series was revised into a digital form with supporting resources, and the publisher provided workshops for teachers on how to use the digital resources. Modern technology has made many mathematical topics more interactive and visual and at the same time, made others automatic and hidden. Allowing graphics and CAS (Computer Algebra System) calculators to be used in external examinations since 2013 has changed the way mathematics is taught and given students with these devices a considerable advantage in terms of assessment.

Looking Forward, Looking Backwards

Looking forward

There are many questions related to mathematics and mathematics education that need to be addressed as we look forward, these include:

- General issues
  - Are subjects (and topics within subjects) an artificial partitioning of knowledge, and would a curriculum that was not subject-based be more appropriate?
  - How might Māori and Pacific Island students' world-views be accommodated? Must we accept that colonialism (and neocolonialism) have 'won the battle' and that western mathematics is more important than traditional cultural perspectives on knowledge?
  - Should learning any subject (e.g., mathematics) be compulsory in schools?
- Mathematics curriculum issues
  - Should relations rather than number be the starting point and the unifying notion in mathematics education?
  - Does emphasizing numeracy distort the mathematics curriculum?
  - How might pure, applied, and statistical mathematics be better integrated?
  - How might mathematical thinking be emphasized more?
  - Is mathematics true or is it a human invention to help us make sense of our world? And, should we be explicit about this with our students?
- Educational issues
  - How can mathematics educators help students develop both group and independent learning skills?

- How might mathematics education shift from being teacher directed to learner driven? (From 'teaching as telling' to 'teaching as asking' and 'learning by doing'?)
- • - Western mathematics education involves listening, reading, and writing; other cultures emphasize visual, practical, and experiential ways of coming to know. How might mathematics education be modified to be more culturally inclusive?
- How might mathematics curriculum reform occur in the future with the increasingly diverse range of people interested and affected by the outcome? (Should it be compulsory? Should it be integrated with other subjects?)

• Assessment issues

- Is the aim of teaching mathematics to have students pass examinations? Tests and examinations teach many students that they cannot do mathematics. Should the emphasis shift from summative to formative assessment?
- How can we shift the emphasis on assessment towards self-assessment?

Looking backwards

Recently there was an unfavourable reaction in the media from numerous teachers and students concerning a mathematics examination organized by the New Zealand Qualification Agency. This reaction has been a catalyst for reflection on the 'progress' of mathematics education in the last 60 years. Knowing the current emphasis on 'back-to-basics' by our government and the Ministry of Education; and the changing background of many teachers (new settlers with very traditional views about both mathematics and mathematics education), the teachers' reactions were somewhat surprising. However, their concern was that the examination was too challenging.

Looking at the examination, it seemed similar in content and difficulty to the high school examinations of 60 years ago; the main difference being that the format was structured to lead the students and to make the assessors' task easier. From a mathematics education perspective there was a much more worrying concern. The examination showed little evidence of the initiatives taken in the last six decades (modern mathematics, mathematical processes, use of technology, statistics, applications, educational aims, thinking, or subject integration). From this perspective one could conclude that the work of mathematics educators has been in vain and no real change has occurred, but hopefully, seeds have been sown which will come to fruition in the future.

References

Anthony, G., & Walshaw, M. (2007). *Effective Pedagogy in Mathematics/Pāngarau.* Wellington, New Zealand: Ministry of Education.

Barton, B. (2008). *The language of mathematics.* New York, NY: Springer.

Barton, B., & The Māori Language Commission. (1991). *Ngā kupu tikanga pāngarau (Mathematics vocabulary).* Wellington, New Zealand: Learning Media.

Begg, A. (Ed). (1991). *Mathematika Pasefika: Vocabulary Database.* Hamilton, New Zealand: Centre for Science and Mathematics Education Research, University of Waikato.

Department of Education. (1900). *The Standards.* Wellington, New Zealand: Government Printer. (Cited by Simon, 1994).

Department of Education. (1969). *Mathematics: Primer 1 to Standard 4. Syllabus for Schools.* Wellington, New Zealand: Department of Education.

Department of Education. (1972). *Mathematics: Forms 1 to 4. Syllabus for Schools.* Wellington, New Zealand: Department of Education.

Department of Education. (1985). *Mathematics: Junior Classes to Standard 4. Syllabus for Schools.* Wellington, New Zealand: Department of Education.

Department of Education. (1987). *Mathematics: Forms 1 to 4. Syllabus for Schools.* Wellington, New Zealand: Department of Education.

Education Review Office. (2016). *Early mathematics: a guide for improving teaching and learning.* Wellington, New Zealand: Education Review Office.

Hunter, J. (Chair). (1998). *Mathematics in New Zealand: past, present and future.* Wellington, New Zealand: Ministry of Research, Science and Technology. (Available from: http://www.msc.vuw.ac.nz/research/morstreport/)

Jones, A. & Jenkins, K. (2016). Bicentenary 2016: The first New Zealand school. *New Zealand Journal of Educational Studies/Te Hautaki Mātai Mātauranga o Aotearoa, 51*(1), 5–18. doi 10.1007/s40841-015-0026-8.

Lewis, D. (1994). *We, the Navigators: the ancient art of landfinding in the Pacific* (2nd edition), Honolulu, HI: University of Hawaii Press. (2nd edition. First published 1972).

Ministry of Education. (1992). *Mathematics in the New Zealand Curriculum.* Wellington, New Zealand: Learning Media.

Ministry of Education. (1993). *The New Zealand Curriculum Framework*. Wellington, New Zealand: Learning Media.

Ministry of Education. (1994). *Pāngarau*. Wellington, New Zealand: Learning Media. [Note: This was a new Māori curriculum, not a translation of the 1992 English-medium one.]

Ministry of Education. (1996). Te Whāriki, He Whāriki Mātauranga o Aotearoa, Early Childhood Curriculum. Wellington, New Zealand: Learning Media.

Ministry of Education. (1999). *Making things count*. New Zealand: Learning Media.

Ministry of Education. (2007). The New Zealand Curriculum for English-medium teaching and learning in years 1–13. Wellington, New Zealand: Learning Media.

Ministry of Education. (2008). *Te Marautanga o Aotearoa*. Wellington, New Zealand. Learning Media.

Ministry of Education. (2009). *Mathematics Standards for years 1 to 8*. Wellington, New Zealand: Learning Media.

Ministry of Education. (2012). *Te Aho Tukutuku - Early mathematics*. Wellington, New Zealand: Learning Media.

Simon, J. (1994). Historical perspectives on education in New Zealand. In: E. Coxon, K. Jenkins, J. Marshall, & L. Massey (Eds.) The politics of learning and teaching in Aotearoa-New Zealand. Palmerston North, New Zealand. Dunmore Press. (pp. 34–81).

Thomas, W. (Chair). (1943). *The post-primary school curriculum*. Wellington, New Zealand. (Report of a committee appointed by the Minister of Education in 1942.)

About the Authors

Andy Begg is an Associate Professor in the School of Education at Auckland University of Technology, New Zealand. He has taught high school mathematics, co-authored mathematics textbooks, and been a Government mathematics curriculum officer. Next, he taught, researched, and supervised postgraduate mathematics education students at the University of Waikato for ten years; continued such work for three years overseas; and then moved to his current position—working with and supervising post-graduate education students.

Jane McChesney is a Senior Lecturer in the School of Teacher Education at the University of Canterbury in Christchurch, New Zealand. She teaches in primary and secondary mathematics initial teacher education and in postgraduate courses in mathematics education and research methods. Her research interests include mathematical practices in early years and school contexts, mathematical tools and representations, and initial teacher education.

Jyoti Jhagroo is a Senior Lecturer in the School of Education at Auckland University of Technology and teaches on undergraduate and postgraduate programmes and supervises postgraduate students. She began her teaching career in South Africa in 1988. Before joining Auckland University of Technology, she taught at a primary school and later at a secondary school as a mathematics teacher in Auckland. In 2012 she completed a research study about the lived experiences of high school immigrant students in mathematics classes.

Chapter III

PAPUA NEW GUINEA: Change and Continuity in Mathematics Education in Papua New Guinea

Kay Owens
Charles Sturt University, Dubbo, NSW, Australia

Philip Clarkson
Australian Catholic University

Chris Owens
Western Sydney University (retired)

Charly Muke
St Theresa College, Abergowrie, QLD, Australia

Introduction

Papua New Guinea (PNG) is comprised of the eastern half of the island of New Guinea and around 600 neighbouring islands, large and small. (The western half of the island was claimed by the Dutch as a colony but is currently controlled by Indonesia.) In the late 1800s, the northern part was designated as a German Protectorate and the southern part as a Protectorate of Queensland (part of Australia) by the British government. Both sections were Protectorate or Territory under the Australian government at the end of World War I. The PNG people had no say in any of these decisions. In 1975, PNG became an independent nation. The government is administered through 22 Provinces. It is very mountainous with some volcanic areas, fast flowing rivers, and wide valleys both in the highlands and in coastal Provinces. The majority of people still rely on agriculture, gathering plants, hunting and fishing even if they have jobs in towns and cities and the majority still live in rural villages, generally with only 20 to 100 houses.

Most remarkable is the diversity of languages, over 850, and hence cultures. These languages are either Austronesian Oceanic or non-Austronesian (Papuan). The Oceanic languages form several clusters. The non-Austronesian are older languages and most are from one large phylum—Trans New Guinea (TNG) phylum which covers most of the main island—and many smaller phyla and isolates. There is more diversity with language used for identity and group cohesion. By contrast, there are similarities among the Oceanic languages with Proto-Oceanic having been formed from these languages established in New Guinea and across the Melanesian Western Pacific nations. All these PNG communities have continued until today as strong identifiable groups, with some continuing changes of land placement occurring even recently due to food shortages, tsunami, relation-building, refugees from West Papua, and disputes.

40 000 to 200 Years Ago

Occupation sites in Papua New Guinea (PNG) have been dated to 40 000 years ago in several Provinces including Central Province. In the Sepik Provinces, evidence shows that trade was made more easily during the minor Ice Age around 10 000 years ago when travel was across an inland lake (Swadling, 2010) and in the Waghi valley of Jiwaka Province, there were drainage systems allowing food production in the valleys at least 10 000 years ago. Despite the number of subphyla, linguists have established a proto-TNG language indicating that these phyla of languages had a long history dating back before the minor ice-age. The other phyla appear to be older.

Later, about 4 000 years ago, the Austronesian Oceania languages developed around the island of New Britain and the obsidian trade was progressing from a neighbouring island, New Ireland, to many other island and coastal areas and even into the mountains of Central Province in the south and into the neighbouring Solomon Islands. Trade or reciprocity routes for other commodities such as shell, nuts, salt and oil developed, often from the coast to the highlands. Trade and exogenous marriages meant that at least some members of different groups became multilingual.

Various aspects of culture such as reciprocity for marriages, warfare, land usage and acquisition, housing, and food management meant that systematic ways of mathematical thinking, associated with spatial and numerical relationships, were required for communication. The languages of the societies hold particular schematic ways of thinking and in many cases this includes different ways of counting. Displays of exchange items and associated rituals would represent numbers and groups of various sizes. Knowledge systems such as designs were also traded and used to establish relationships.

Mathematical systems grow and change as they meet a society's needs. Hence it is no surprise that there are numeral systems in the ancient PNG cultures (K. Owens, Lean, with Paraide, & Muke, 2017). Thus the 2-cycle and (2, 5) cycle, (2, 5, 20) cycle (mostly digit tally), and body-part tally numeral systems with multiple variations in Australia and New Guinea are part of the cultural knowledge. Language, DNA, and archaeology indicate that these existed before the city-states of the Middle East were being established. There are, for example, clusters of body-part tally systems whereas some languages of the East Sepik-Ramu Subphylum developed various 4-cycle systems. Two other subphyla include some languages with 6-cycle systems which appear to have developed through local trading. Similarly, the 10-cycle system entered the New Guinea region with the original Austronesian immigrants around 4 000 years ago.

Within the village, decision making often occurred in the men's houses. Decisions were made in the Austronesian as well as the non-Austronesian languages. Most village activities involved mathematical thinking. Mathematical activities included navigating across the seas to land and islands beyond the horizon, making canoes, drums, ceremonial seats, and houses (some large communal long houses, men's houses, yam houses, and sleeping houses that were round and some rectangular, some on posts, some covered with kunai that looked like a 'stone' from the distance, some over the water, some with hand-made planks, some with woven walls). Designs were modified to meet environmental and ecological situations. Mathematical designs changed with contact (Paraide, 2010, 2016). For example, coastal houses often took up Samoan ideas from Samoan missionaries while highland houses have had the planks replaced by woven walls. These new designs were then taught to other young men through involvement in the mathematical activity from a young age. It seems that these examples are simply the latest in a practice that has lasted for many thousands of years.

Gardens were also made with due consideration to need. In some places, there were carefully spaced drains and mounds providing area patterns. Nevertheless, only length measures were used in negotiations, the visuospatial reasoning of size being observed and valorised (Owens, 2015). Similarly, exchanges of pigs were by eye or length measures as well as counting. Other traditional mathematical understandings were around topology especially in the making of continuous string bags and other objects (*bilums*). More recently, materials for these bags have changed and new designs created and swapped. Some patterns spread from women to other women, across thousands of kilometres and many languages but generally as part of a relationship. Similarly, the geometric designs of *kapkap*, a sign of leadership, have changed and been traded (K. Owens, 2015; Were, 2010).

Late 1800s to Early 1900s

European exploratory expeditions gathered pace in the late 1800s and then came settlements. The Cambridge Expeditions in the 1890s saw counting systems recorded along with other linguistic and cultural commentaries (e.g., Ray, 1895, 1907, 1912). At this time colonising settlements were founded on the Huon Peninsula in the Morobe and Madang Provinces (Wagner & Reiner, 1986), New Britain (Threlfall, 2012) and in Central Province around Port Moresby (Legge, 1956). In the settlements, schools were quickly set up by the various Church organisations. In particular, the Methodist in East New Britain (e.g., Danks' 1878 papers on his ENB students and 1880 school for chiefs on Duke of York Island). The report of 1886 noted that about 30 students under a particularly well-regarded Tongan missionary teacher had learnt to read and do a little arithmetic but all sang beautifully. The London Missionary Society (LMS) had schools along the Papuan coast while the Catholics had schools in the New Guinea islands as well as west of Port Moresby, and in a few other places. The Lutheran Churches began education at Finschhafen, Morobe Province, in 1888 and they attempted to learn and use the local language. Their education system spread in Morobe and Madang areas, reaching into the highlands. Not long after this, the Anglicans (Church of England) set up schools in Oro (Northern) Province. There were varying attitudes to the Indigenous cultural practices but all saw education as important in changing beliefs especially those like sorcery that impacted in awful ways on people as well as seeing education as a way forward in the world's approach to education such as learning to read and write (Gash, Hookey, Lacey, & Whittaker, 1975). A concern of the LMS was for protection of the Papuans rather than political rule.

Australian Territories Period: After 1918

At the conclusion of the First World War, the League of Nations (forerunner of the UN) handed the German colony to Australia as the Trust Territory of New Guinea. By then the British had formally handed the Papua colony to Australia to administer. Although officially governed separately, there was a clear synergy in how Australia administered New Guinea and Papua. (For more documents on this period see Gash et al. (1975).)

As Australia assumed formal responsibility for government, the Catholic and Lutheran churches spread extensively in the New Guinea Trust Territory with the Catholics also active in Papua. In all areas of church work, they were establishing primary schools as well as churches. Church workers were using local languages in church services and teaching literacy with some mathematics. These early colonial schools thus established mathematics as a school phenomenon that later became a Western English phenomenon. More specifically in New Guinea, the Lutheran church had Tok Ples (local language) schools or they used one of the lingua franca: Yabim (along the coast), Kôte (in the mountains) with Bel along the Madang coast. (Some Tok Ples schools have continued to today as part of a cooperative supported by Summer Institute of

Linguistics (SIL) to teach literacy and numeracy at the pre-elementary level.) In Papua, Police Motu became a lingua franca and was used in many government areas although English predominated in schools.

However, at the Treaty of Versailles and the handing over of the mandated territory of New Guinea to Australia, the military administration of New Guinea, concerned about the number of 'enemy' among the expatriate population including the missions, closed the German administration schools until a civilian administration set up a system of formal education for both government and missions in 1921. The boys at the government school at Kokopo in East New Britain subsequently came from many provinces and their first teacher W. C. Groves (1923) claimed that this assisted in breaking down inter-tribal hostilities and he noted how he was gradually learning about the psychology of his learners to their educational benefit. He noted the value of understanding more about cultural psychology for teaching purposes if they were to enhance the students' knowledge. Another teacher noted that more emphasis should be placed on agriculture rather than other technical subjects (Smith, 1987). In 1946, Groves became the first Director of Education of Papua and New Guinea.

Rural stations, generally run by an Australian Patrol Officer with a support force of local people, had a school to which children came from surrounding villages (up to several days' walk away) and often having different languages. They were taught in English or a local lingua franca if the teacher knew it (often they came from further away).

The Queensland Inspector of Education in 1930 came and examined 1118 Papuan students and 80 'mixed race' and placed schools into three groups: those schools taught by white teachers and generally satisfactory and progressive; schools under white missionaries with 'native' teachers with disadvantages when the missionaries were absent on patrol duties; and those solely taught by 'native' teachers who produced 'very satisfactory mechanical results' for Standards 1 and 2 (Grades 1 and 2) but lack of English made it difficult for upper grades. Satisfactory schools received grants for teaching. However, Percy Chatterton began a year's training for these teachers and proficiency was raised (Smith, 1987). European children had their own schools. In the Methodist schools of

Papua and New Guinea Islands, many of the missionary teachers were Pacific Islanders. As early as 1929, the government anthropologist F. Williams began a paper for the Papuans, called *The Papuan Villager*, to talk about lives of Papuans from different places and ensuring multiple cultural links were encouraged in schools.

For the schools with a technical emphasis, some basic mathematics was involved in record keeping and dealing with money as well as measurements. The missions ran these programs and the mission schools were inspected by government officials. The government also provided expectations for Standards 1 and 2. For Standard 2 it stated for arithmetic:

- Mental: Mental exercise in the four simple rules and easy money calculations.
- Tables: Multiplication tables to 12 times 12. Money tables. Measures of length, area, weight (avoirdupois only), and of time.
- Written: The four simple rules including long division, compound addition, subtraction, multiplication and division. Notation and numeration up to 100 000.
- Practical application of the foot rule in measurement of yards, feet, inches and halves, fourths, and eighths of an inch. (Smith, 1987, p. 59)

Some important aspects of education were the emphasis on technical education supported by the churches and administration. For example, Hubert Murray (1920) explained that this would support diligence and self-sufficiency in the economy so that villagers could run their own plantations and maintenance industries rather than serve white Australians and others. However, Murray also noted that the mission-trained mechanics and so on were not coming to town to be available for the government services (Smith, 1987). This situation may have begun the strong attitude that schooling should ensure employment outside of village life, an issue which continues to this day together with many concerns about school drop-outs not staying in villages.

In 1945, Wedgwood noted that lighting, overcrowding, sitting on floors or in desks that are too high, and multi-grades made teaching difficult (Smith, 1987). By 1938, there were five Standards and 77 students examined at the highest level out of 2473 students.

Some larger towns, like Goroka, became magnets for people, especially when a road system was developed. Gradually in such towns, the Creole Tok Pisin began to dominate in every day conversations. Interestingly, English became the language of instruction for many schools, although in the lower grades local languages and/or a lingua franca could often be heard.

The question of education beyond primary school and in tertiary institutions was beginning to be considered with different opinions about the viability of such institutions due to colonial attitudes during Murray's administration (Smith, 1987). Similarly, there was consideration about cultures and the blending of cultures. Williams wrote in 1935:

> The complacent conviction that our own culture is at all points superior to the native's, and the idea that it may be transplanted by simple straightforward conversion, has led us to force European traditions and learning upon him without regard to their fitness or usefulness. It is only within comparatively recent years that we have recognised that the methods, and the very subjects, used and taught in European schoolrooms are not immediately suitable for the native pupil; and the 'education for life' in his case is education for a life which is at many points radically different from the European. (Williams, 1935, p. 2)

The government began schools for Europeans in 1930s but not for the Chinese or mixed race children who relied on missions or schools run by the Chinese Nationalist Party. In 1927 there were 34 168 children enrolled in mission schools mostly on the islands and coastal areas. Boarding schools were available and near Madang some English was taught as required by the government but students were mostly learning to read and write one of the local lingua franca *Bel* so they could communicate between themselves as they came from quite different languages. There were some conflicts between the different missions although the government tried to allocate specific areas to each one of the churches and several had teachers who were well versed as pastors. These men also travelled beyond the frontiers of Australian control, bringing early

Christian education and some literacy and numeracy. Training of teachers for government schools also continued in small numbers (Smith, 1987).

In 1932 there was a request from the League of Nations for opinions about the language of instruction to be collated. The missions and governments had different views: the missions wanted the vernacular to be used given they also wanted to minister to their beliefs and behaviours and the Lutherans provided a detailed submission, but it suggested that the local languages were simpler than English and more appropriate to their 'minds'. There were still fears about the role of education in the re-positioning of 'natives' when compared with the white settlers in the workforce. Griffith's policy suggested the missions set up the schools and supply the teachers for village school, primary or middle school and high school with the vernacular or lingua franca of the district used in the village school, and the lingua franca, not English in the primary school and for the 6 years of high school English would be the medium of instruction. However, the new administrator McNichol did not proceed entirely with this plan as he set up new administration schools (Smith, 1987). English was the medium of instruction with Tok Pisin used outside in the gardens and games. At the government school in East New Britain, a Solomon Islander who could speak the East New Britain language (Kuanua or Tolai) was head, some girls were enrolled and selected students learnt about the wireless and semaphore (Smith, 1987). Later Groves, having studied anthropology, returned to recommend in 1936 that schooling appropriate for village life rather than a European curriculum be provided (Smith, 1987). Further developments of the education system were delayed by the eruption of the volcano in Rabaul and the setting up of the administration in Lae in Morobe Province on the mainland. The dominance of German among the expatriates, Tok Pisin among the locals, and lack of training meant that McNichol did not find a workforce to teach elementary arithmetic, geography or English to his standards. Despite increased revenue from gold mining in Morobe, there was no significant increase in the education budget compared with health, agriculture and public works (Smith, 1987).

Groves and the other Australian education administrators continued to favour village schools teaching in the vernacular in the 1950s supported by broadcasts in vernacular, Police Motu, Tok Pisin and English. Vocational training and agriculture were also seen as important venues for promoting education for all and not for an elite (Smith, 1987). There were Area Education Centres to support these features for adults and the community as a whole.

Both Territories saw front line fighting during the Second World War and much of the northern islands, northern coastal regions and northern coast of Papua were occupied by the Japanese. There were a few schools such as in Wewak where the first Chief Minister went to school that taught in Japanese but most schools, being mission schools, were closed despite the efforts of some New Guineans to continue them, as they were considered by the Japanese to be subversive. In southern Papuan coastal areas, teachers continued as best they could with little or no European supervision and virtually no resources but still children became literate in their language and simple arithmetic (Wedgwood cited in Smith, 1987). The recruitment of all young men and leaders from along the Papuan coast to assist the Australian army (Wedgwood cited in Smith, 1987) resulted in them perishing or otherwise not continuing with schooling. There was a hole left in the fabric of these communities (Temu, 2012).

However, the war also provided the impetus for the government (then called Australian New Guinea Administrative Unit - ANGAU) to establish a Central Papuan Training School to train teachers, medical orderlies, and artisans. Wedgwood then proposed a 30-year plan for education after the war to ensure there was not just an educated elite nor a perception of a second rate education but that the Territories have appropriately trained directors, funds, and a linguistic survey to ensure vernaculars were recorded for 'native' education (often using the recordings of missionaries) (Smith, 1987). In some areas, such as Oro Province where the Japanese invaded, there was enormous loss of teachers and ravages of war to overcome. However, the administrator Murray also emphasised the importance of the welfare and education of the Papuans and New Guineans and pressed for funds and personnel to ensure this happened but progress was hindered by his administrative inexperience and lack of detailed guidelines in Ward's directive for the Territory. The health budget was three times the education budget. The plan for training for those who served in the war, including

Papua New Guineans, does not seem to have been completely fulfilled but some did receive this higher education.

At that time, Soldier Settlements (small farming blocks) were given to Australian ex-servicemen in the two Territories as was done in Australia. As in Australia, these blocks were not sustainable but they did lead to a small influx of Australians in the Territories. During the later part of Papua and New Guinea Territories, Australia administered both T-school (Territory) and A-school (Australian) primary schools with dual curricula, often on the same school grounds. The T-schools were staffed by both Nationals and Australians (or other expatriates) but the latter were gradually phased out. The national teachers had generally obtained a two-year teaching qualification, post Year 9 or 10 and at the time most primary school teachers from Australia were also two-year trained ex-Year 11, the last year of school. A-schools occurred in all cities and large towns and in some more rural areas where there were a significant number of Australian or expatriate children. There was not much difference in the mathematics curricula in the two types of primary schools.

Another delayed consequence of the war has been the aid provided by Japan, mainly for infrastructure projects and building computing expertise in the country. Together with Australia and New Zealand, Japan continues to provide funding for education projects in PNG.

Groves' notion of schools or areas being able to plan their own curricula for the needs of their community was overturned by the Minister for Territories in Canberra, Australia, P. Hasluck. One reason for this was that the young Australian teachers did not change practices to adapt their teaching to Papua and New Guinea's context (Smith, 1987). One commentary noted than an arithmetic problem read

> An orchardist plants 1289 trees, 487 are apple trees, 395 are peach trees, and the rest are plum trees. How many plum trees are there? (Roscoe, 1958, p. 95)

The question was asked without regard to the unlikelihood of such large numbers of fruit trees being in the village and that these varieties do not grow in Papua New Guinea.

Hasluck set out to ensure money was spent on government and mission schools, teacher training, and technical training. Papuan and other teachers were being registered. English was to be a key for education but Tok Pisin was seen as having a clear mandate for general communication.

The number of girls receiving education in schools was well below boys at this stage, and there were different curricula for them. The diversity of schools and the need for registration indicated that mission teacher training would qualify a teacher for the village school to Standard 2. The elementary level required Standard 9 while the Intermediate level required two further years at a teacher training facility.

By the mid-1950s, scholarships to do secondary education in Australia were created, with mixed responses by the Australians and others in the Territories. The concern was that the Australian government was moving too fast or losing the blend of cultures that the earlier anthropology-trained educators had suggested. However, during the 1960s, there was still concern about an elite being formed, the disparity between the regions, the educational expansion being insufficient for economic or social independence or self-sufficiency, and the lack of local economic development to absorb school leavers appropriately. Among these changes was the stratification of three T-schools depending on contact with English and the classification of A-schools for students with English as a mother tongue. With the 1955 memorandum from Hasluck on universal primary education with writing and reading in English, G. T. Roscoe detailed a plan for this to be achieved by 1975 (cited in Smith, 1987, pp. 211–214). A similar call for universal education was made in the late 1980s, and again in 2012 but it should be noted that this is far from being achieved in 2016. The main reason is the doubling of the population with a large increase in young people (National Statistical Office, 2014).

While European colonies were being given independence in Africa and elsewhere, the Dutch in West Papua were providing Papuans' secondary education to localise the public service. At that time, the principal of the Australian School of Pacific Administration (ASOPA), J. Kerr, encouraged an educated elite in the Papua and New Guinea Territories. In 1961, a legislative council met for the first time with elected as well as appointed members including John Guise who was later to become the Governor General at Independence in 1975. Guise (1961) called for universal village education to Grade IV, and secondary in English taught by Australian and Papuan teachers together with a university college associated with a Sydney university, and more grants-in-aid to local government councils and missions for education.

With the report of the United Nations Visiting Mission to New Guinea (1962), known as the Foot Report, the provision of a university education as well as the administrative college was now strongly recommended. Ken McKinnon, who was Director of Education, showed that the number of students in secondary schooling would reach 13 600 by 1968 (with 7 750 in government schools and 5 850 in church or non-government schools). Some students in Form 4 at two of the government schools were able to achieve an Australian examination result, and scholarships continued providing students with insight also into Australian culture. Thus education for an elite occurred. The Teachers' College at Goroka was set up along with the University of Papua New Guinea (UPNG). To avoid delay, the first preliminary year began at the University in March 1966 in the Port Moresby Showground with overcrowded dormitories and an inadequate study centre several kilometres away. Teacher education was needed for the mushrooming primary schools but it was also seen as a stepping stone for other employment. The Administrative College that was set up in 1963 moved to Waigani, near UPNG in 1968. This emphasis on secondary and tertiary education reduced funds for primary education, resulting in a slower rate of increase in numbers beginning education. However, by 1967, there were nearly 200 000 students enrolled in primary education (McKinnon, 1968). There were more than a thousand expatriate (mostly Australian) teachers in both the primary and secondary schools with most primary schools phasing out the overseas teachers by Self-Government in 1973. While teachers were recruited from Australia, often studying also at ASOPA, some of the male teachers received six months' training in Rabaul learning how to teach English and the lower grades of primary school and then taught in more remote locations. Both government and church technical education expanded in villages and towns. Not only was education seen as a political and social need, but the World Bank, most urban elite, rural newcomers to education and administrators saw education as important economically (Smith, 1987). However, the reality was that economic benefits did not come immediately to individuals and families from their investment in education. The 1966 Census showed 44% were subsistence farmers with an additional 37% mostly subsistence farmers and/or hunter-gatherers.

Leading up to Independence 1975 until 1990

Australian administrators in the Territories and politicians began to recognise that Papua and New Guinea needed independence. In 1973, there began a preparation period of self-governance with Papua and New Guineans taking leading roles in education and other portfolios. Independence was achieved in 1975. However, it is useful in this section to comment on the changes that were taking place in education some 10 years or so prior to Independence, and then from Independence to the early 1990s. This 25 to 30-year span saw the country, and education in particular, assert its independence in a very real way.

The privileged adopted attitudes of elitism and there became a distinct divide between the elite and non-elite that is exacerbated today by the 'big man' mentality that is also associated with successful money making, including through gambling. In particular those with degrees are expected to give and to have an entourage of supporters, especially in highland cultures. This is particularly the case for those with doctorates (Brown, 1990; Pickles, 2013).

The Five National Goals set out in 1973 in planning for Independence in 1975 were titled Integral Human Development, Equality and Participation, National Sovereignty and Self Reliance, and Conservation of Natural Resources and Papua New Guinean Ways. Some of the Eight Point Principles for achieving the Five National Goals aimed to improve social justice, reversing some of the implementation of the above mentioned Foot Report with its acceptance of an elite. A quota system of selection from each school rather than the pure meritocratic exam system meant that a certain percentage could be considered from each school together with consideration of other backgrounds. More girls were encouraged to go to school. Much of the control of education had been devolved to Provinces but national curriculum remained. The Eight Point plan proved to be more rhetoric than reality (King, Lee, & Warakai, 1985). It was extended by the Matane Report (1986) with its emphasis on integral human development rather than manpower needs. This report focused on the social, vocational, spiritual, and political growth of individuals and communities and brought about the biggest change in education (Weeks, 1993) but was implemented with restructures (National Department of Education Papua New Guinea, 1991, 1995) that were not easily deliverable (National Department of Education Papua New Guinea, 1993), and were still not fully addressed at the time of writing in 2016.

The school system

At Independence, the primary schools became known as Community Schools with the intention that the school would service the community and be supported by the community. The teaching staff for Community Schools was nationalised. Many schools were run by churches but were funded mostly and controlled by the government. By 1990, the schools had Community School Boards and funding came through the Provinces. However, some Australians or expatriates continued to teach in the high schools with mostly expatriate teachers in the senior high schools through the 1980s. The level of English of teachers and hence of students was high.

At Independence, the A-schools were reduced in number and became fee-paying International Education Agency (IEA) schools. By the early 2000, most of the teachers in the IEA schools were nationals. The PNG government provided oversight and approval of curricula and some were supported, especially for high school subjects by Distance Education from NSW, Australia. Non-Papua New Guinean children were required to attend these schools and a growing number of children of Papua New Guinea professionals or business men also attended. In 1997, IEA became an autonomous group providing curricula and professional development.

Secondary education had not been a high priority for the Australian government. Hence at Independence there were relatively few high schools compared with the number of Community Schools, although in the prior 10 years before Independence this had started to change. This reflected on-going debates about universal education and elite education (Legge, 1956). Secondary education was two tiered: Provincial High Schools taught Grades 7 through 10 with at least two per Province. Hence most students had to leave their own village and board to attend a Provincial High School. Years 11 and 12 were taught in four National High Schools, which were wholly boarding schools. There was a deliberate policy to mix students from different regions in each of the National High Schools (NHS) in an attempt to ensure the future leaders of the country would come to see PNG as one country rather than an amalgam of different tribal regions.

Three key policies that were inherited from the Australian government were kept at Independence. The first was the assessment regime. There were external tests given at the completion of Grades 6, 8 and 10. These and the families' inability to pay the modest school fees meant that there was an attrition rate of about 50% for each of these years. The second policy dealt with the language of teaching. The colonial government had deemed English as the language of teaching for all grade levels in the 1960s, against much opposition from religious missionary schools. There are various accounts in the literature of children who began school at about age 7, and in some cases for a year or more were not able to understand what the teacher was saying but got along by imitating their peers who for some reason had some understanding of English (Clarkson, 1991a; Muke, 2012). Third, the PNG government retained the notion that all schools, including religious schools, would teach from the one common curriculum if they wished to receive government money to support their endeavours. Again, although there was some opposition, virtually all schools accepted this proposition, and those that did not at independence, soon joined the fold.

The universities

The University of Papua New Guinea (UPNG) was established in Port Moresby in 1965 with mainly expatriate staff. However, soon after Independence, the newly appointed Head of Mathematics was a Papua New Guinean. Gradually more PNG staff were recruited, often the best of the teachers from the school system. Many of these were able to study abroad and obtained higher degrees. In the Department of Mathematics there was a clear research emphasis on pure mathematics.

An offshoot of the Faculty of Education was the Education Research Unit. Members of the Unit carried out important research into education that had an impact on the school system. Some of the early studies concentrated on Piagetian ideas and showed that PNG students reached Piaget's formal level later than overseas (Shea, 1978). Cross-cultural studies were used to investigate the relevance of Piagetian stages in other cultures. However, others critiqued this testing in terms of language and culture. Later the Unit and Faculty cooperated with the National Department of Education (NDoE) in the *Indigenous Mathematics Project* (see later). Major contributions in critiquing the education system were made by Sheldon Weeks and Gerard Guthrie (2011).

In 1968, the PNG Institute of Technology moved from the shadow of UPNG in Port Moresby to PNG's second city, Lae in the north, and became the PNG University of Technology (UNITECH). With a dedicated group of expatriate staff, the Department of Mathematics set up service mathematics courses for the other Departments of the University. In the early days most students beginning their studies at UNITECH did so with a Grade 10 entry and the Mathematics Department provided them with bridging mathematics courses depending on in which Faculty they were enrolled. Some reports indicated this year prepared students well for University with less attrition or failure than the often younger NHS students (C. Owens, 1979). PhD degrees were awarded from 1974 onwards at UPNG and established early in UNITECH's life. This privilege was not achieved at the University of Goroka until 2010.

In the mid-1970s a three-year *Mathematics Education Project* (MEP) began with support from the Nuffield Foundation and Monash University in Australia. The aim of the MEP was to develop appropriate approaches for teaching mathematics to the early students at UNITECH by undertaking cognitive research on specific mathematics topics and taking account of their different cultural backgrounds. Researcher teachers were recruited from the UK, New Zealand and Australia. Lesley Booth particularly researched and prepared materials on algebra, later influencing Australian programs. Thus the first editions of the self-paced individualised modules were developed for first and second year students (Wilkins, 2000). The MEP provided concrete materials that were regularly used to introduce complex mathematical ideas together with practical problems tailored for each of the technology degrees, an approach that was later strengthened through the use of computer-supported modelling (see below).

With the success of the MEP, moves were made to establish a permanent Centre as an adjunct to the Department of Mathematics. With a substantial grant from the British Council, the Mathematics Education Centre (MEC) was founded with Allen Edwards, from England, as the Director. Although the MEC continued to coordinate and print the Mathematics Departments booklets for the various mathematics units, the Department staff took over the writing of new booklets and revision of older ones. A crucial aim for the MEC became the support of entry-level students to the university and hence it began work with the school system by providing some professional development and liaison with the NDoE staff.

At that time Alan Bishop from Cambridge University, England, and then Ken Clements from Monash University, Australia each spent three months at the MEC working closely with Glen Lean in the Department. Alan's visit stimulated his interest in the impact of culture on mathematics (Clarkson, 2008), which culminated in his seminal work on mathematics enculturation (Bishop, 1988) and the companion volume on enculturation or transitions (Bishop, 2002; de Abreu, Bishop, & Presmeg, 2002). With Glen Lean, both Alan and Ken furthered their particular interests in visuospatial reasoning in mathematics, but now using a PNG lens. There was a series of influential individual and joint publications, to which others (Owens and Clarkson in particular) later added (Bishop, 1978a, 1978b, 1979, 1983; Lean, 1984; Lean & Clements, 1981; K. Owens, 2015). Bishop's visit to PNG was a turning point in his influence on mathematics education, resulting in the award of the International Commission on Mathematics Instruction's Felix Klein Award in 2016. Two years after the MEC was founded, Philip Clarkson was appointed Director. The Centre continued its support of the Department of Mathematics and work with schools, but there was a decided shift to fulfil the third aim, which was to focus on research (Clarkson, 1983b). An early move was to appoint two Research Officers, Naomi Wilkins (who went on to hold senior administrative positions at UNITECH for many years) and Wilfred Kaleva (who went on to a PhD in ethnomathematics with Bishop, Director of the Glen Lean Ethnomathematics Centre (GLEC), Head of Department of Mathematics, and Associate Dean of Research at the University Goroka).

Mathematics Heads of Department of NHS, and representatives of University mathematics institutions formed an NHS examination committee. Among other roles the MEC Director was appointed as the external examiner for Grades 11 and 12 mathematics. A process was instigated where the four Mathematics Heads of Departments of the National High Schools (NHS), and one mathematics lecturer from each of the UPNG, UNITECH and the College of Distance Education met mid-year to draw up suitable items that could be used on the final external mathematics papers. After a couple of years there was quite a worthwhile item bank available. It also allowed a certain collegiality between the Heads to grow, given their schools were many flying hours distance from each other. After the mid-year meeting the Director (Committee Chair) drew up the final examination papers. Following the end of year examinations, the group was again hosted at UNITECH to mark the completed papers and then review the outcomes. A final report on each year's process was published as an MEC Technical Report. This process formed a template that the NDoE subsequently used for examining most subjects. The Director and other university staff continued to provide professional development for all school levels, the teachers' colleges, and various government departments, including nursing (Clarkson & Sullivan, 1981). The Director became a member of the Advisory Committees of all school level curricula. Several of the expatriate staff continued with research and supporting education in PNG; in particular Pak Yung, Phil Clarkson, Peter Sullivan and Kay Owens (see below). Many staff from UNITECH became professors or influenced education overseas.

As to the research program of the MEC, regular conferences were held and overseas visitors always stimulated new thinking. These included Kath Hart, Alan Bell and Brian Wilson from England, Beth Southwell (on three occasions) and Gilah Leder from Australia, David Lancy from the USA and Claude Gaulin from Canada. Annual research conferences were run by the Centre, with books of reviewed papers published for each conference (Clarkson, 1981b, 1982, 1983a, 1984). The conferences attracted participants from across PNG but also from Fiji, Samoa, Tonga, Solomon Islands, and various islands in Micronesia. As well as papers given by visitors, it was also an opportunity for staff at UNITECH to report on research they were undertaking. During this time the Centre published 23 research reports and 13 technical research reports, adding to the four previously published research reports.

The UNITECH Mathematics Department was an early user of computer technology with a mainframe and two computer laboratories in the early 1980s which were soon followed by desktop computer laboratories. Statistics and modelling were particularly strong aspects of the Department with lecturers such as Peter Jones, Terry Fairclough, John Lynch, Ian Wright, Chris Wilkins, and Don Lewis. The country's engineers, architects and natural resource developers continue to have a high level of mathematics although Wilkins (2000) considered it was not as high as Australia and other places overseas.

With the arrival of Professor Kathleen Collard in the mid-1980s, there was a decided move away from the mastery booklets and towards a more traditional style of university teaching. However, she also encouraged problem-solving as a basis of mathematics learning and this was also incorporated into the teaching booklets. This extended the group work that had developed around the content of the booklets. By then, most entry students were Grade 12 graduates. Professor Collard initiated the Diploma in Engineering Mathematics that was considered equivalent to the third year of a British university's mathematics degree. After several PNG staff completed this Diploma she was able to use her influence to ensure they obtained entry to Masters and PhDs overseas. Among the early graduates was Samuel Kopamu who went on to lecture at UPNG, UNITECH, to be Head of Department and a senior academic at the University of Goroka. John Gesa was another graduate and lecturer at UNITECH.

During all this time, Glen Lean was working on the most remarkable collation and analysis of PNG's many counting systems. For 22 years his students completed questionnaires on their own counting systems. From these and other written records, often from first contact, he recorded counting systems of two-thirds of the then-considered 750 (now 850) languages. By using the cycle system analysis he produced a monumental documentation of the origins of counting systems in PNG and Oceania (Lean, 1992; K. Owens, 2001; K. Owens et al., 2017).

Peter Sullivan worked on nursing mathematics after the Centre was approached by the hospital in Lae to help with the mathematical difficulties that their nurses were experiencing (Sullivan, 1981; Sullivan & Clarkson, 1982). He subsequently completed his M.Phil. from UNITECH on this issue. Owens continued this research by producing a number of well-regarded, frequently republished sets of inservice teaching materials targeted at specific nursing and health worker contexts (K. Owens, 1986–1987). Other studies considered attitudes (Clarkson & Whippy, 1981), language (Dube & Siegel, 1981), problem solving (Dube, 1981) and performance (Jones & Wilkins, 1981; Wilson, 1981).

Although producing a number of research projects dealing with cross-cultural issues (Clarkson, 1984b; Clarkson & Leder, 1984), Clarkson's main research project dealt with the impact of language on mathematics learning. He began by establishing that there was a linkage (Clarkson, 1983c, 1984c), and then spent a number of years starting to tease out the relationship (Clarkson, 1991a, 1991b, 1992a, 1992b; Clarkson & Clarkson, 1993; Clarkson & Galbraith, 1992; Clarkson & Kaleva, 1993). Of this research, it was later noted that these "results suggested that bilingual students who were competent in their home language and the language of teaching outperformed other students in mathematics, even when other factors such as socio-economic status and parental education were accounted for. . . . (they also) showed that the notion of bilingualism had to be far more nuanced than had been the case in earlier research, and global applications of deficit models were just not appropriate . . . the ways teachers and schools dealt with students' multiple languages in relation to mathematics learning was for the most part an unrecognized but critical issue in both research and practice" in mathematics (Barwell et al., 2016, p. 7). Charly Muke continued this line of research some years later while completing his doctoral studies with Clarkson (Muke, 2012;

Muke & Clarkson, 2011a, 2011b). Clarkson continues to work on the interrelationship between language and mathematics learning.

In some ways, it was an historical accident that a Mathematics Education Research Centre was founded at UNITECH, clearly a technological university, and not at UPNG which had a Faculty of Education. The MEC certainly met its initial aim of developing and printing a suite of booklets for use in teaching mathematics in the early years of the university programs. But as the quality of entry students improved and the Mathematics Department moved from being merely a service department and began programs in mathematics for its own students, budget issues saw the Centre cut. Sadly, the role it played in providing professional development for teachers was not picked up elsewhere in the wider system. Research in mathematics education to inform policy and curriculum in PNG was partially taken up by the National Research Institute, especially with the work of Patricia Paraide, with some research emanating from the GLEC Centre University of Goroka.

The mathematics curriculum

At Independence, as it had been since the early 1960s, the school mathematics curriculum was little different from Australian State systems. This was a period around the world of New Maths. Thus the textbooks for the children were written with all the set theory symbols making it even harder for the children with English as a second language to grasp the basic concepts. Furthermore, each day of the week covered a different topic such as number, geometry or measurement. There was little consolidation or meaning.

Significant too was a visit by Zoltan Dienes who brought his multi-base arithmetic blocks to schools and advised on TEMLAB (Territory Mathematics Laboratory). After all, there were many different language groups that used cycle systems other than 10. However, these proved to be difficult for children, working outside of their own cycle system for additions and subtractions just as in other parts of the world. (Subsequently, by the 1990s, these objectives were dropped from the syllabuses for primary schools.) Beth Southwell (1974) completed a doctorate on mathematics education in Papua New Guinea, becoming the first Australian woman with a doctorate in mathematics education. She looked at the development and implementation of TEMLAB curriculum and materials that began in 1968 and was still being implemented in 1974, too long for on-going development and teacher education. TEMLAB consisted of box of cards and much equipment; and required teacher education, teacher organisation and group work to be effective if cards or materials were not to be lost or misordered. Furthermore, TEMLAB in PNG showed that the New Maths symbolic and language registers were inappropriate for second language learners in PNG and that without adequate training teachers were not confident and resorted to rote teaching. TEMLAB, it was hoped, would reduce the rote learning and increase the investigation in mathematics. Although TEMLAB did not continue, it did, however, set the scene for continuing use of concrete materials. It continues to be difficult to distribute materials for teaching in PNG and provide adequate training for teachers. Nevertheless, as all future curriculum documents encouraged the use of learner materials, especially those readily available in the village, it also highlighted the difficulty of distributing materials and teacher inservice learning to schools. The short teacher education courses also encouraged group work and concrete materials for learning although with increasing numbers in classrooms at school and at colleges and universities, this was hard to maintain. Some teachers at least and some College lecturers realised that the one traditional way of teaching mathematics was not the only possibility: there were alternatives.

At that time also, PNG females did not perform as well as males, indicating that opportunities for female school children and teachers needed to be promoted (Southwell, 1974). This remained an issue for education, noted by important reports such as Matane's and future Australian Aid projects. Despite considerable efforts to promote gender equity, it remains elusive.

In the second half of the 1970s, the National Department of Education took a dual pronged approach with principal researchers and curriculum writers working together. The researchers set up the *Indigenous*

Mathematics Project (IMP) to explore PNG mathematics coordinated by David Lancy (Principal Research Officer in the NDoE). This took an ethnomathematics approach (Lancy, 1978, 1979, 1981) and prepared a special edition of the PNG Journal of Education (1978). References were made to different ways of counting, understanding algebra, using spatial reasoning, and the difficulty of expressing measurement ideas in local languages. At the same time, Geoffrey Saxe was undertaking his world renowned investigation by exploring the mathematics of ordinary life with the Oksapmin communities in the remote west of PNG and deliberating on the interplay between culture and cognitive processes (Saxe, 1982, 2012, n. d.). Souviney (1983) particularly considered the impact of textbooks in five different classroom areas. This wealth of ideas influenced the mathematics curriculum writers in PNG for many years.

At the same time, Murray Britt headed-up the second prong of the initiative. He led a team mainly of PNG nationals. They rethought the Grades 7 and 8 mathematics curriculum and the way it was taught, and developed textbooks for these years using the title *Mathematics Our Way*. The title certainly emphasised the nationalistic spirit very evident post-Independence. As well as drawing on the ethnomathematics promoted by the IMP and hence trying to use local contexts and materials in teaching mathematics, these books also promoted problem solving, one of the other emphases that had impacted the teaching of mathematics worldwide (Mathematics Our Way, 1981; Roberts, 1982; Souviney, 1981).

The IMP project was a breakthrough in understanding that the mathematics embedded in the life of ordinary PNG people was in many ways different from western mathematics that had been taught in PNG schools. It had become obvious that the mathematics taught in schools was school maths, but that had little to do with your village life. Hence, even though Britt and his team utilised 'everyday contexts' in the textbooks, and urged teachers in the accompanying teacher handbooks to find examples from village life to enrich their mathematics lessons; teachers, students or parents were finding it was hard to bridge the gap. Either the new textbooks did not arrive or the approach was rejected in the schools although there was some professional development in the lower grades with TEMLAB. After some years with few or no replacement books available, the fewer and fewer tattered books in schools often remained on shelves as teachers returned to the chalkboard. Interestingly, the *Maths Our Way* series was discontinued and textbooks for Grades 9 and 10 were published as *Secondary School Mathematics* in 1984 and 1985, produced by the PNG national team that Britt had established, and using the same principles. The Grades 9 and 10 books were accepted with little upset but many of the activity based exercises were excluded, based on trials. The Grades 7 and 8 *Maths Our Way* texts were being phased out by 1989 with a return to traditional mathematics teaching with a focus on overseas linearly organised and abstract curricula returning, with only a hint of exercises reflecting names of people and activities relevant to PNG society. Only gradually, reference to mathematical thinking relevant to PNG activities continued to some extent with the influence of University of Goroka lecturers interested in ethnomathematics, such as Kaleva, Matang, and Bino.

In the early 1980s, the curriculum writing team also turned its attention to the Community School mathematics curriculum. Interestingly, they started with the mid years and produced books for Grades 4, 5 and 6 titled *Community School Mathematics*. Hence they were able to make interesting connections with the books already published for the Provincial High Schools. Building on some notions brought to the fore by the IMP project, the writing team again tried to build in with examples that were set in PNG village life. However, there were few real excursions into ethnomathematics. Nor was there much attention paid to the language and mathematics research being carried out in the MEC, although this was being aired by then in various forums. There was however some effort to use problem solving ideas in these books. Interestingly, the texts were not printed in PNG, but in Australia by Oxford University Press.

The resultant texts were colourful and noticeably had pictures of objects without English, to be counted in local languages but the notions of ethnomathematics had little impact on the writers. The teachers' guides directed teachers through the books and to some extent encouraged teachers to take account of PNG context. However, there was only the one place in the teachers' handbook in Grade 1 that dealt with using non-English languages in teaching. Like previous published texts, after a few years, the large classes of 45 in most schools had only a few textbooks to share in the class. Sadly this experiment of outsourcing to a multi-national company for the writing, production and printing of the series that in the end covered all Grades 1 to 8 left the NDoE with a huge debt.

Across the grades and years, textbooks continued to be replaced, new ones written and produced with new titles and huge costs, but often schools went without materials and teachers without professional development. Often those whose voices were heard wanted content that was western or too advanced, especially at primary levels.

Teacher education

There were two government teachers' colleges in Port Moresby and Madang, with multiple church-affiliated teachers' colleges. Three of these were supported by different Protestant churches and three by Catholic orders and churches with the Seventh Day Adventist (SDA) church training teachers in Fiji. Gradually the colleges were filled with just Grade 12 graduates. There were advisory boards for national curricula for school subjects and for related subjects at the teachers' colleges. These boards drafted and revised curriculum and often endorsed textbooks and other materials. In the teachers' colleges, subjects (courses) covered basic mathematical skills as well as mathematics education. The course increased to a three-year diploma, but for financial reasons this course was delivered over two years.

Community School Teachers' Colleges were offering first two-year certificates and then three-year diplomas, but to be done in two years for cost purposes. This was a strain on the staff and students. There was a National Curriculum and inspection from the Office of Higher Education. By 1990, most mathematics staff were nationals, usually trained as high school teachers. Port Moresby Teachers' College became the Papua New Guinea Institute of Education. The latter was mostly providing inservice training to raise teachers with certificates to diplomats (a role that all the colleges had in the 1970s). It later provided training for elementary school teachers, mostly by establishing units of work that students finished after attending a short course at a regional centre, run by a staff member. When

Provinces managed education, the training of elementary teachers was taken by a coordinator and district trainers. Few of these were trained in early childhood or studied mathematics education or cultural mathematics.

PNG high school teachers mostly began their careers with a three-year Diploma from Goroka Teachers' College (GTC). Some then went to UPNG for a two-year course leading to a Bachelor degree (B. Ed inservice) while a few did the four-year B. Ed program at UPNG. GTC was founded in 1961 originally as a teachers' college for primary schools, but in 1968 it became a second avenue for secondary teacher education. In 1975 Goroka became part of UPNG with some coordination between the two in terms of UPNG converting Diplomas to Degrees. Many of the mathematics staff such as Rod Selden and David Shield went on to lecture at UNITECH. L. Ross (1982) continued with research on school mathematics standards; and Dave Godwin returned to Australia.

Changes from 1990

During the late 1980s and through the 1990s, major reports were produced at government level that had a profound impact on the education system. Underlying the main thrust of this movement was for PNG to build a world class education system, but not lose contact with the thousands year old traditions that marked PNG life out as being quite unique.

The school system

Central to the changes that occurred in the school system was the report produced by Paulus Matane (1986). Matane called for universal education in PNG. He was concerned for women's equity and moral and education standards, and continued to encourage mathematics to maintain PNG ways. He took note of World Bank reports that suggested a restructuring of the school system so all children could attend school and most receive an education to Grade 8. Through World Bank finances and quality overseas advisers, the system changed, but not in all places.

The structure of the school system did change in the 1990s. Village elementary schools with locally recruited teachers were to teach Pre-elementary and Elementary 1 and 2. Primary schools would teach from

Grades 3 to 8, and the Provincial High Schools, whose numbers expanded, taught Grades 9 to 12. Interestingly, the four National High Schools (Grades 11 and 12) remained. At the same time, more and more administrative tasks such as teachers' pay, inspections, decisions on schools, and professional development were devolved to the Provinces.

In the flush of enthusiasm for the change in the system, hundreds of elementary schools were built and supported by the village communities. But this mushrooming of schools was not adequately supported in terms of training of teachers and provision of resources. The elementary school teachers were expected to have Grade 10 and to speak their local language so they could teach in their local language. However, it was hard to register the schools and the teachers, which was a necessary step in teachers receiving their government pay. With only two-weeks a year of training, followed by self-instruction units, these teachers received inadequate education in how young children learn arithmetic. They also were not adequately prepared to teach in their village language or to bridge to English in Grade 2, and hence many reverted to the manner in which they had been taught, often using Tok Pisin, or English if they themselves spoke English regularly, and hence defeating the main aim of why elementary schools had become part of the system.

Primary schools also had their own difficulties. Grade 3, the first year of the school, was deemed to be a follow-up 'bridging year'. It was anticipated that students entering the school would have completed their elementary schooling in their village language and attained basic arithmetic concepts with some bilingual work in Grade 2; and hence Grade 3 became a bilingual or multilingual teaching year although culture and language were expected to be used throughout the primary school. However, few teachers accepted this proposition and believed that English should be the main language of teaching in all years, falling back to the use of Tok Pisin or a village language when needed (Muke, 2012).

Even with these reforms, testing continued to impact on the progress of many students to complete schooling beyond Grades 6, 8 or 10. Little change had occurred in the attrition rates at these levels. The issue of whether to keep an elite group and managing the system of schools for more and more students required examinations to exclude children. There was continuing unrest about the value of going to school or not with so many disincentives such as lack of on-going schooling or opportunity to get a job.

The curriculum

The mathematics curriculum continued to undergo change from the 1990s, which made it difficult for teachers. Before changes that flowed from the Matane Report came to fruition, the development of mathematics textbooks continued, but now starting at Grade 3. In 1989 there were also commissioned textbooks, taking this out of the hands of the writing team that had built up expertise in the NDoE. The exception was the elementary school curriculum writers who produced teachers' guides and support materials for teachers, although these were frequently unavailable in elementary schools.

By the late 1990s there was a detailed curriculum reform for the whole of the school system. The *Curriculum Reform Implementation Program* (CRIP) introduced an outcomes-based education for schools, rather than an objective-based approach. This followed international trends. The most difficult aspect of this reform was the restriction on syllabus materials supplied to schools. The outcomes were not elaborated upon and so teachers were not easily able to plan well to meet these outcomes, especially when textbooks or teachers' guides were unavailable. New textbooks emerged, still with exercises relevant to PNG.

Matane's request for mathematics to take account of culture was not really implemented in this externally advised and very brief syllabus. Multilingual education was not built into the syllabus. In fact, the documents assumed that much of the teaching would be in English, even in bridging in Elementary 2 and Grade 3. The mathematics curriculum assumed English would be the language of teaching. In practice, the language was generally Tok Pisin or an English with multiple dysfunctions in terms of words and grammar. Although there was some use of ethnomathematics ideas, they were not a driving force in the documents. In terms of mathematics, however, children who had the opportunity to learn some mathematics in their vernacular (Tok Ples) in elementary school achieved better in early arithmetic tasks than those who did not have this opportunity, although children in town schools (presumably with some English in the home) achieved slightly more with English as the only language of instruction (Matang & Owens, 2006, 2014; Paraide, 2003).

Teacher education

In 1995 all UPNG education programs were unified at the Goroka Campus of UPNG and in 1997 the Campus was constituted as the University of Goroka (UoG). It was offering education degrees for preservice and inservice teachers. UoG gradually expanded its offerings into Technical Vocational Education, Agriculture, Health, and Business Administration. With the increase in high schools throughout the country there were more teachers needed. UoG tried to meet this demand but with intake rapidly doubling to what it had been, sadly resourcing did not keep pace with this demand.

Two other developments at UoG are worth noting. As noted above, since the late 1970s there had been a growing recognition of the role ethnomathematics could play in school mathematics teaching. Although over the years there was a variable emphasis on this in the curriculum and textbooks, nevertheless it persisted. This concept was strengthened by significant members of the PNG mathematics education community obtaining higher degrees in studying the implications of ethnomathematics in PNG. These included Wilfred Kaleva (1998) PhD studying with Alan Bishop; Rex Matang (1996) MEd and a nearly complete PhD—both long term staff members at UoG; Charly Muke; Frances Kari (1998) PhD also studied with Bishop and later taught in Distance Education at UPNG; Patricia Paraide, PhD (2010). At present, the relevant department at UoG is now named the Division of Mathematics Computing and Ethno-mathematics. The current (2016) Vice-Chancellor Musawe Sinebare was Head of Department when the Glen Lean Ethnomathematics Centre began in 2000 with some of Glen Lean's materials sent to PNG by Alan Bishop. Wilfred Kaleva requested Kay and Chris Owens catalogue his papers and write about his work and unpublished thesis (Owens, 2001).

In 2001 the Glen Lean Ethnomathematics Centre was officially opened by Geoffrey Saxe when Rex Matang became Director. Kay has continued to work with UoG staff on research. Lean's materials were digitised (USA National Science Foundation grant through Pacific Resources in Education and Learning (PREL), Hawaii). The Centre has produced further research (e.g., Matang & Owens, 2014) and promoted interest in the mathematics of PNG (K. Owens, 2016). In particular, UoG staff have continued to encourage an ethnomathematical approach in curricula and in teacher education, especially through the elective *Mathematics, Language and Culture*. The teachers' colleges had a similar elective but it was not often taught. However, at least some class time in all courses was expected to establish an awareness of PNG mathematics and its value for school mathematics.

With the school reforms being enacted, a comprehensive program was put in place in 1998 to improve the quality of the preservice education programs offered in the teachers' colleges and at UoG. This was the *Primary and Secondary Teacher Education Program* (PASTEP) led by Steve Pickford (2008). The program operated in five of the then six teachers' colleges and UoG. The focus was on revising the college programs and to work with the college staff to initiate modern trends in teacher education such as collaborative inquiry group work, functional grammar, problem solving and contextual mathematics, gender equity and other ethical approaches, and multi-grade planning. UoG staff visited universities in Australia but PASTEP made little progress in this institution. In addition, computer laboratories for each college with staff trained in ICT also came to fruition, particularly so they could communicate between colleges and to provide electronic resources for staff and students along with library books supplied to each college. Each overseas adviser worked long-term with a counter-part. Overall, the results seemed to show that both practices and content in teachers' colleges improved with PASTEP. The Program brought in a number of overseas experts to work in the colleges for varying lengths of time. Each expert was paired with a PNG national for the expert to have an immediate local to turn to for advice, and for the national counterpart to learn from the expert. There is no doubt that in virtually all cases the national counterparts excelled in their learning during this time. The Program held yearly conferences for college staff, when the staff members had to present ideas and products that they had been working on during the past year. These proved highly beneficial not just for the sharing of ideas but the camaraderie that it started to build across colleges.

However, there was some criticism of the PASTEP, the most cogent of which was that yet again western ideas were being imported for the teachers of PNG and there was not enough, although some, emphasis on building a PNG way of teaching (Nongkas, 2007). In an evaluation of PASTEP it became evident that many college staff members struggled financially with their family expectations. Student attitudes, large classes, and less dedicated staff were not supportive of the efforts of dedicated staff. Most mathematics lecturers in the colleges had little idea of ethnomathematics, believed that there was little influence of language on mathematics learning, and thought that the emphasis on Grade 3 being a bridging year was not applicable to mathematics. The same results seemed to apply to early year teachers already employed in schools. Although not related to PASTEP directly, the evaluation of final year college students revealed that their knowledge of mathematics (and English) was not of a level that built confidence that school mathematics would improve, although some excellent teaching was observed (Clarkson, Hamadi, Kaleva, Owens, & Toomey, 2004).

Just as PASTEP was finishing in 2003, college staff, who might have been looking forward to a time of consolidation, had to cope with another crisis. Because of budgetary considerations it had been deemed necessary to continue teaching the content of the three-year diploma, but squeeze it into a two-year time frame. On a brighter note, some 40 teachers' college lecturers took the opportunity to obtain their Masters in Education degree through a *Virtual Colombo Plan*, an Australian Aid program taught at UoG. The opportunity for this Plan was not repeated but an alternative was made available on one further occasion which enable more staff to upgrade their qualifications. These staff included a number of the colleges' mathematics staff. However, a new wave of lecturers were not always managing to understand and use the materials provided by the PASTEP project, partly due to later local attempts to upgrade revisions of the college courses.

Throughout all the changes that were occurring in the lower year levels, the standards at the National High Schools remained high. With such a high attrition rate throughout the lower school years, clearly only the better students were reaching Grades 11 and 12. However, it was a fact that the NDoE Mathematics Curriculum Advisory Committee constantly kept the standard equivalent to higher levels of the Australian high schools but the curriculum was not tied to the Australian curricula in any way. Another check on quality was the involvement of university staff in the setting and external marking of the final examinations. The quality of staff in the National High Schools was high. However, most Grade 12 students were now coming from overcrowded Provincial High Schools.

Throughout all the years, the Summer Institute of Linguistics (SIL) provided word lists, some in-depth studies of mathematical ideas, and numeracy education with their literacy education. They have continued to assist curriculum officers, trainers, and some teachers in remote areas on good practices for teaching literacy (especially in the vernacular language of the children) and provided a teachers' guide for elementary school mathematics to assist the limited outcomes-based syllabus.

Only Madang Teachers' College remains a government college, with some now affiliated with the universities. The PNG Institute of Education became an inservice college where teachers mostly went to upgrade from a Certificate to a Diploma. Later it provided courses for the elementary schools and ran some longer courses on campus. By 2000, most institutions were staffed by 98–100% nationals with mainly one or two volunteer expatriate staff, some who had become nationals but some locally recruited and had lived in PNG for some time.

PhDs were highly valued at the universities but the workloads were too high for continuing research, and accountability and management of funding was still a struggle. One of the most difficult aspects of education in PNG in the 21st century was the doubling of students at the University of Goroka wishing to become secondary teachers, but without a proportional increase in academic and professional staff. Many students unable to get into secondary school were doing Distance Education with varying degrees of success.

The Technical Colleges continued to teach mathematics for trades and other technological situations. However, there were issues with equipment and so mechanics might be taught just from a diagram. UoG also provided Technical Vocational Education.

Some felt the mathematics being taught at the University of Goroka and teachers' colleges was appropriate for teacher education, others did not. It was hard to recruit overseas staff as the salaries were much lower than previously. Local staff struggled to make ends meet with their family expectations. Dedicated staff were not always supported by students able or willing to attend and do their best as the large classes made it difficult especially in general education lectures.

The National Department of Education has continually gathered data on achievements, from national testing. There is evidence that students entering primary schools have not achieved well in mathematics at elementary schools. Some blame the lack of teaching in English, others the outcomes-based rather than objectives-based earlier curriculum but the real issues appear to be the lack of teacher education, the pay for teachers, and the limited syllabus.

New universities

From the late 1990s another interesting development was the founding of the Divine Word University (DWU) supported by the Catholic Church which has grown to be a significant provider in the education field. DWU took over the Catholic Teachers' College in Wewak which became the University's Faculty of Education. The Catholic Teachers' College in Mount Hagen is also affiliated. DWU is having an influence in mathematics and mathematics education in PNG. An early doctorate, Lakoa Fitina taught mathematical computing and is now at UPNG. Cecilia Nembou, who has a doctorate in operations research, has had considerable influence not only in mathematics but also in administration and leadership. In addition, Patricia Paraide now works in the postgraduate and research section of DWU, having been at the forefront of research in mathematics education while at the PNG National Research Institute. Her study of Tolai mathematics education (Paraide, 2010) and bilingual education for mathematical conceptual and cultural development is significant (Paraide, 2014).

Vudal Agricultural College became the PNG University of Natural Resources and the Environment in East New Britain. The Pacific Adventist University also began outside Port Moresby. Two Catholic teachers' colleges in Wewak and Mt Hagen aligned with DWU while Madang TC attempted to amalgamate with UoG. A Lutheran University with a seminary and teachers' college was also planned.

After 2013

Due to the extraordinary degree of poor numeracy and literacy in English, the next important policy decision on education in 2012 by the Prime Minister, P. O'Neil, reversed many decisions including the language of instruction (Paraide, 2014), outcomes-based education (National Department of Education, 2003), and school structure. English was to be the language of instruction. There was to be free education in the hopes that the schools would recruit all children (this has led to even larger class sizes). Standards were introduced incorporating behavioural objectives. The new mathematics syllabus has been supported by Japanese advisors. While the syllabus may not be easily understood by all teachers, at least at the elementary school level, the syllabus gives ideas around topics to be covered, performance indicators and assessment. Teachers' guides encourage the same thing to be taught in English in every school with partially scripted single page lessons. Today all teachers in all schools, curriculum officers and administrators are Papua New Guinean except for an occasional volunteer.

Changes in teacher education

Around 2001, a Diploma to train the elementary teachers was prepared and it was going to be implemented in both government teachers' colleges and others but this did not eventuate. By 2010, early childhood education courses began to develop at the University of Goroka. Some early childhood centres catered for the growing number of working mothers but also some developed in the rural and remote areas as part of the stimulus of the University of Goroka staff. PNG was beginning to see an increase in private education institutions from early childhood through to university. Teachers' colleges began in several other provinces and some of these have been registered. Another university college began in Port Moresby.

Issues for Education and Mathematics Education

The latest education policies overturned the 1990's commitment for elementary school classes to be taught in the language of the village and Grades 2 and 3 to be bridging years. That decision had been arrived at after years of careful consideration of education research completed in PNG, and research beyond PNG that was applicable to its context. That decision has been called into question and has prompted a call for far more dialogue between researchers, tertiary staff, teachers and politicians (Clarkson, 2016; Paraide, 2014). The 1990s reform and principles were inadequately implemented in terms of teacher education at the lowest level, continuing professional development, and management of salaries and resources.

School mathematics in PNG has gone through an enormity of changes in the last 50 years. There have been many changes of official curricula and a number of attempts at building a robust foundation for students to be able to live their lives with dignity. Clearly more money is needed for the fundamental learning aids that teachers can use in their classrooms. Most of all there needs to be a long period of stability, albeit with well thought through incremental changes as the system finds its own way to an authentic PNG way of good mathematics teaching.

References

Bishop, A. (1978a). Spatial abilities and mathematics in Papua New Guinea. *Papua New Guinea Journal of Education. Special edition Indigenous Mathematics Project, 14,* 176–204.

Bishop, A. (1978b). Spatial abilities in a Papua New Guinea context (Vol. Report 2, Mathematics Education Centre). Lae: Papua New Guinea: University of Technology.

Bishop, A. (1979). Visualising and mathematics in a pre-technological culture. *Educational Studies in Mathematics, 10*(2), 135–146.

Bishop, A. (1983). Space and geometry. In R. Lesh & M. Landau (Eds.), *Acquisition of mathematics concepts and processes* (pp. 176–204). New York: Academic Press.

Brown, P. (1990). Big man, past and present: Model, person, hero, legend *Ethnology, 29*(2), 97–115.

Clarkson, P. (2016). The intertwining of politics and mathematics teaching in Papua New Guinea. In A. Halai & P. Clarkson (Eds.), *Teaching and learning matheamtics in multilingual classrooms* (pp. 43–55): Sense Publishers.

Clarkson, P., & Whippy, H. (1981). Papua New Guinea university students' attitudes to mathematics. In P. Clarkson (Ed.), *Research in Mathematics Education in Papua New Guinea 1981* (pp. 21–37). Lae, PNG: PNG University of Technology, Mathematics Education Centre.

Dube, L. (1981). Research on the teaching of problem solving at PNGUT: Retrospective and prospective. In P. Clarkson (Ed.), *Research in mathematics education in Papua New Guinea 1981* (pp. 38–42). Lae, PNG: PNG University of Technology, Mathematics Education Centre.

Dube, L., & Siegel, J. (1981). Teaching the English of mathematics: Does it make a difference in the learning of mathematics? *International Journal of Mathematical Education in Science & Technology, 12*(4), 449–452.

Gash, N., Hookey, J., Lacey, R., & Whittaker, J. (1975). *Documents and readings in New Guinea History: Prehistory to 1889*. Milton, Queensland, Australia: Javaranda Press.

Groves, W. (1923). Report for the first years of school *Groves Papers*. Held at University of Papua New Guinea.

Guise, J. (1961). Maiden speech by John Guise. *Territory of Papua New Guinea Legislative Council Debates, 4*(1), 11 April.

Jones, P., & Wilkins, C. (1981). Mathematics achievement at tertiary level: A comparative study at PNGUT. *Papua New Guinea Journal of Education, 17*(1), 72–81.

King, P., Lee, W., & Warakai, V. (Eds.). (1985). *From rhetoric to reality: Papua New Guinea's Eight Point Plan and National Goals after a decade*. Waigani, Port Moresby, Papua New Guinea: University of Papua New Guinea Press.

Lancy, D. (1978). Indigenous mathematics systems: An introduction. *Papua New Guinea Journal of Education, 14*(Special issue), 6–15.

Lancy, D. (1979). The Indigenous Mathematics Project - 1977–1980 *Education Research - 1976–1979: Reports and essays* (pp. 45–54). Konedobu Papua New Guinea: Standards Division, Provincial Education Services, Ministry of Education, Science and Culture.

Lancy, D. (1981). Indigenous Mathematics Project - an overview. *Educational Studies in Mathematics, 12*, 445–453.

Lean, G. (1984). *The conquest of space: A review of the research literatures pertaining to the development of spatial abilities underlying an understanding of 3-D geometry.*

Paper presented at the Fifth International Congress on Mathematical Education, Adelaide, Australia.

Lean, G. (1992). *Counting systems of Papua New Guinea and Oceania.* (Unpublished PhD Thesis), PNG University of Technology, Lae, Papua New Guinea.

Lean, G., & Clements, M. (1981). Spatial ability, visual imagery, and mathematical performance. *Educational Studies in Mathematics, 12,* 267–299.

Legge, J. D. (1956). Australian colonial policy: A survey of native administration and European development in Papua. Sydney: Angus and Robertson.

Matane, P. (1986). *A philosophy of education for Papua New Guinea.* Waigani, Port Moresby, Papua New Guinea: Government Printer.

Matang, R., & Owens, K. (2006). Rich transitions from Indigenous counting systems to English arithmetic strategies: Implications for mathematics education in Papua New Guinea. In F. Favilli (Ed.), *Ethnomathematics and mathematics education, Proceedings of the 10th International Congress on Mathematical Education Discussion Group 15 Ethnomathematics.* Pisa, Italy: Tipografia Editrice Pisana.

Matang, R., & Owens, K. (2014). The role of Indigenous traditional counting systems in children's development of numerical cognition: Results from a study in Papua New Guinea. *Mathematics Education Research Journal, 26*(3), 531–553. doi: 10.1007/s13394-013-0115-2

Mathematics Our Way. (1981). *Papua New Guinea Education Gazette, 15*(10), 270–273.

McKinnon, K. (1968). Education in Papua and New Guinea: The twenty post-war years. *The Australian Journal of Education, 12*(1), 8–9.

Murray, J. H. P. (1920). Review of the Australian administration in Papua. Port Moresby, Papua New Guinea: Government Printer.

National Department of Education Papua New Guinea. (1991). *Education sector review.* Port Moresby, PNG: Author.

National Department of Education Papua New Guinea. (1993). *The education reform.* Port Moresby, PNG: Author.

National Department of Education Papua New Guinea. (1995). *National education plan.* Port Moreseby, PNG: Author.

National Department of Education, P. N. G. (2003). *National curriculum statement for Papua New Guinea.* Waigani, Port Moresby, PNG: Author.

National Statistical Office. (2014). 2011 National population and housing census. Port Moresby, PNG: National Statistical Office, PNG.

Owens, C. (1979). A subsequent survey of student progress at the Papua New Guinea University of Technology *Report No. 2 Department of Chemical Technology* Lae, PNG: PNG University of Technology.

Owens, K. (1986–1987). *Drug calculations, Ordering drugs, Solutions, Maternal health mathematics.* Port Moresby: Department of Health.

Owens, K. (2001). The work of Glendon Lean on the counting systems of Papua New Guinea and Oceania. *Mathematics Education Research Journal, 13*(1), 47–71.

Owens, K. (2015). Visuospatial reasoning: An ecocultural perspective for space, geometry and measurement education. New York: Springer.

Owens, K. (2016). Culture at the forefront of mathematics research at the University of Goroka: The Glen Lean Ethnomathematics Centre. *South Pacific Journal of Pure and Applied Mathematics, 2*(1).

Owens, K., Lean, G., with Paraide, P., & Muke, C. (2017). *The history of number: Perspective from Papua New Guinea and Oceania.* New York, NY: Springer.

Paraide, P. (2003). *What skills have they mastered?* Paper presented at The National Education Reform: Where Now, Where to?, Goroka, Papua New Guinea.

Paraide, P. (2010). Integrating Indigenous and Western mathematical knowledge in PNG early schooling. (PhD), Deakin University, Geelong, Victoria, Australia.

Paraide, P. (2014). Challenges with the implementation of vernacular and bilingual education in Papua New Guinea. *Contemporary PNG Studies: DWU Research Journal, 21*(November), 44–57.

Paraide, P. (2016). Chapter 11: Indigenous and western knowledge. In K. Owens, & Lean, G. (Ed.), *History of number: Evidence from Papua New Guinea and Oceania.* New York, NY: Springer.

Pickles, A. (2013). *The pattern changes changes: Gambling value in highland Papua New Guinea.* (PhD), University of St Andrews, UK. Retrieved from https://research-repository.st-andrews.ac.uk/bitstream/10023/3389/6/AnthonyPicklesPhDThesis.pdf

Ray, S. (1895). The languages of British New Guinea. *The Journal of the Anthropological Institute of Great Britain and Ireland, 24,* 15–39. doi: 10.2307/2842475

Ray, S. (1907). *Reports of the Cambridge anthropological expedition to Torres Straits* (Vol. 3). Cambridge: Cambridge University Press.

Ray, S. (1912). Notes on the languages in the east of Netherlands New Guinea. In A. Wollaston (Ed.), *Pygmies and Papuans: The stone age today in Dutch New Guinea* (pp. 322–345). Smith Elder: London, UK.

Roberts, B. (1982). Community school mathematics: The international connection. *Papua New Guinea Journal of Education, 18*(1), 88–94.

Roscoe, G. (1958). The problems of the curriculum in Papua and New Guinea. *South Pacific,* 95–96.

Ross, L. (1982). The effects of standards in mathematics if national high schools are doubled. In D. Harvey & W. Palmer (Eds.), *Seminar on educational standards* (pp. 46–52). Goroka, PNG: Goroka Teachers College.

Saxe, G. (1982). Developing forms of arithmetical thought among the Oksapmin of Papua New Guinea. *Developmental Psychology, 18*(4), 583–594.

Saxe, G. (2012). Cultural development of mathematical ideas: Papua New Guinea studies. New York: Cambridge University Press.

Saxe, G. (n.d.). Cultural development of mathematical ideas. from http://www.culturecognition.com/

Shea, J. (1978). The study of cognitive development in PNG. *Papua New Guinea Journal of Education, 14*(Special Issue: Indigenous Mathematics Project), 85–112.

Smith, P. (1987). Education and colonial control in Papua New Guinea: A documentary history. Melbourne, Australia: Longman Chesire.

Southwell, E. (1974). A study of mathematics in Papua New Guinea. In The Australian College of Education (Ed.), *Educational perspectives in Papua New Guinea*. Canberra, Australia: Editor.

Souviney, R. (1981). Teaching and learning mathematics in the community schools of Papua New Guinea (Education, Trans.) *Indigenous Mathematics Project Working Paper Series* (Vol. 20). Port Moresby: Department of Education

Souviney, R. (1983). Mathematics achievement, language and cognitive development: Classroom practices in Papua New Guinea. *An International Journal, 14*(2), 183–212. doi: 10.1007/BF00303685

Sullivan, P. (1981). Investigation into the mathematics of nurses-in-training in PNG. In P. Clarkson (Ed.), *Mathematics education research in Papua New Guinea 1981* (pp. 121–128). Lae, PNG: PNG University of Technology, Mathematics Education Centre.

Sullivan, P., & Clarkson, P. (1982). Practical aspects of drug calculations (Vol. 24). Lae, PNG: PNG University of Technology, Mathematics Education Centre.

Swadling, P. (2010). The impact of a dynamic environmental past on trade routes and language distributions in the lower-middle Sepik. In J. Bowden, N. Himmelmann, & M. Ross (Eds.), *A journey through Austronesian and Papuan linguistic and cultural space: Papers in honour of Andrew Pawley* (Vol. 615, pp. 141–159). Canberra, Australia: Pacific Linguistics, ANU.

Temu, D. (2012). *A lone Papuan voice and the myth of the so-called Fuzzy Wuzzy Angel.* Paper presented at the Kokoda: Beyond the legend, Australian War Memorial.

Threlfall, N. (2012). *Mangroves, coconuts and frangipani: The story of Rabaul.* Printed by Gosford City Council, NSW, Australia: The Rabaul Historical Society and Neville Threlfall.

United Nations Visiting Mission to New Guinea. (1962). Report. 21st session, United Nations Trusteeship Council. New York, NY.

Wagner, H., & Reiner, H. (Eds.). (1986). *The Lutheran Church in Papua New Guinea: The first hundred years 1886–1986.* Adelaide, Australia: Lutheran Publishing House.

Weeks, S. (1993). Education in Papua New Guinea 1973–1993: The late-development effect? *Comparative Education, 29*(3), 261–273.

Were, G. (2010). *Lines that connect: Rethinking pattern and mind in the Pacific.* Honolulu, HA: University of Hawai'i Press.

Wilkins, C. (2000). Department of Mathematics and Computer Science, History. from http://www.unitech.ac.pg/InformationAbout/Departments/MatheMatics-ComputerScience/History

Williams, F. (1935). The blending of the cultures: An essay on the aims of native education. Port Moresby, Papua New Guinea: Government Printer.

Wilson, C. (1981). The performance of National High Students at University. In P. Clarkson (Ed.), *Research in mathematics education in Papua New Guinea 1981* (pp. 146–153). Lae, PNG: PNG University of Technology, Mathematics Education Centre.

About the Authors

Dr Kay Owens began her teaching career as a mathematics and health education secondary teacher in Australia before moving to Papua New Guinea for 15 years with her partner Chris Owens. She taught mathematics at the Papua New Guinea University of Technology and taught, as Head of Department, health education and education at Balob Teachers College. On returning to Australia, she taught mathematics education at the now Western Sydney University for 15 years before moving to be with family in Dubbo to Charles Sturt University for 14 years. She was Vice-President (Publications) for the Mathematics Education Research Group of Australasia and held numerous positions in State and regional professional groups. She is a 35-year Member of Australian College of Educators following her work in PNG, and assists with mathematics and environmental associations. During her years in Australia she has continued to work with Papua New Guinea colleagues in joint research projects on ecocultural mathematics and mathematics education. She has numerous published papers and two of her books focus on Papua New Guinea—*Visuospatial Reasoning: An Ecocultural Perspective for Space, Geometry and Measurement Education* and *History of Number: Evidence from Papua New Guinea and Oceania.*

Before joining Australian Catholic University, **Emeritus Professor Philip Clarkson** for nearly five years was Director of a Research Centre at the Papua New Guinea University of Technology. During this time, he carried out research into language and mathematics education and is well known internationally for this work; Charly Muke became one of his doctoral students. Prior to that Philip was at Monash University and tertiary colleges in Melbourne. He began his professional life as a teacher of chemistry, environmental science, mathematics and physical education in secondary schools. Philip Clarkson has led major consultancies and ARC research projects, was President, Secretary and Vice President (Publications) of the Mathematics Education Research Group of Australasia, foundation editor of *Mathematics Education Research Journal*, served on various editorial boards of both professional and research journals, and published widely. He continues to supervise research students, speak at various international conferences both in mathematics and science education, gives workshops for teachers and publishes regularly.

Chris Owens carried out research at the University of Sydney in physical chemistry and in analytical chemistry at the University of NSW. In 1973 he began his 15 years working at the PNG University of Technology in Lae rising to Acting Head in the Department of Chemical Technology. During his time in PNG, he completed an MSc in Chemical Education from the University of East Anglia, UK and a BEdStud(P/G) from the University of Queensland. Returning to Sydney in 1988 he accepted a lectureship in the Faculty of Science, Food and Horticulture at the now Western Sydney University where he worked for 15 years until retiring in 2003. He continues with some casual teaching and tutoring including mathematics. In 2006 he assisted with a review of the year 11 and 12 science courses in PNG. Professionally he was heavily involved with the PNG Institute of Chemistry and the Australian Institute of Mining and Metallurgy (PNG Branch). He is a Fellow of the Royal Australian Chemical Institute and was for some years Chair of its NSW Chemical Education Group.

Dr Charly Muke began his studies in mathematics and mathematics education at the University of Papua New Guinea and completed further studies at the University of Goroka. He was a mathematics teacher for many years before taking up a position at the University of Goroka and then a teaching position at St Theresa's College, Abergowrie, QLD, Australia. He received a research Masters qualification from Waikato University, New Zealand, for his work on his Mid-Wahgi counting system and he is co-author of a chapter in the book *History of Number: Evidence from Papua New Guinea and Oceania*. He completed his PhD on the use of languages in collaborative group work in Papua New Guinea schools. He set up a library foundation in Jiwaka Province and continues to support the schools and education system in that Province bringing crates of books to the schools from his Australian contacts. He was also involved with Elementary Schools mathematics in the Province bringing workshops to teachers and providing the teachers with computers with the professional inservice materials. This was part of an Australian Development Research Award with co-investigators Kay Owens and Vagi Bino from University of Goroka and others.

Chapter IV

AUSTRALIA: Mathematics and its Teaching in Australia

Judith A. Mousley
Deakin University

Chris Matthews
Griffith University

Early Australia: Our Shared History

Most Australian histories start with the arrival of James Cook in 1770 and the subsequent "first fleet" in 1788. But long before the establishment of Athens and Babylon—at least 60 000 years ago—Australian Aboriginal peoples had developed a sophisticated culture with, for example, its own versions of education, medicine, astronomy, agriculture, and other fields. Australian Aboriginal peoples are the world's oldest continuing cultures[1]. Australia was colonised by England under the doctrine of *terra nullius*, that is, empty land. Indigenous people were considered to be a primitive race that would eventually die out to make way for the "civilised", more technologically advanced culture of the newcomers. As a consequence, Indigenous people were portrayed, consistently and erroneously, "back home" in England as savages with no educational methods or mathematical knowledge. Such suppositions are still repeated today by teachers and other influential people, although scientists are leading the

---

[1] Archaeological evidence suggests a population of 750 000 Indigenous peoples in 1778, although some estimates are for more than one million persons (Australian Bureau of Statistics, 2017).

questioning of such assumptions. For example, Glen Lean (1994) reported the outcomes of his 22-year study of Indigenous counting and measuring systems of Oceania, Polynesia, and Melanesia as well as ways that these became part of the Indigenous culture in Australia for thousands of years prior to European settlement. Complex concepts of measurement, space, time, and direction were all based on extensive mathematical thinking that was different from the "western" thinking of colonists. Lean noted that the European colonial times led to the decline of many Indigenous mathematical systems (Lean, 1994, Abstract).

A colonial heritage

Captain Cook was not the first Westerner to land on the continent: the French, Dutch, Spanish, and Portuguese had all mapped various parts of the continent. However, Cook claimed it for England in 1770.

At that time, the French were showing interest in the South Pacific and America was refusing to accept more convicts, so English settlement of Australia commenced with the colony of New South Wales [NSW][2] being a penal settlement. Convict fleets arrived at Port Jackson and Norfolk Island (with both settlements being called "Sydney") in 1788, with more fleets arriving in 1790 and 1791. There were 26 school-age children of these "first settlers". Free settlers arrived in 1793, but the Colony of NSW remained essentially penal until 1823.

By 1780 some "dame schools" had been established in tents or rough huts (Burkhardt, 2014).

> [Dame schools are a] stage where the teachers have a very poor general education and little or no professional preparation, where the syllabus is sketchy and poorly defined and the emphasis is on completely mechanical, rote performance in the three R's . . . associated with poor standards of attainment. (Hughes, 1969, p. 131)

The majority of teachers during the first two decades of the colony were convicts or ex-convicts, but by 1800 a few free settlers received

---

[2] NSW originally covered the area of what are now the states of New South Wales, Victoria, and Queensland, South Australia and the Northern Territory.

salaries from the "Society for the Propagation of the Gospel in Foreign Parts", whose name signaled the main purpose of schooling as well as the key curriculum content of the time. With further convict fleets settling in Norfolk Island, Lieutenant Governor King appointed the colony's first experienced teacher, for boys only.

> The boys are taught in the town of Sydney (Norfolk Island) by [convict] Thomas Macqueen, who was once a schoolmaster in England and has merited by his good conduct the opinion of the Governor. (Marsden, 1796, in Burkhardt, 2014, p. 2)

This catering for "white" male children was a pre-cursor of educational inequity for years to come.

Inequalities in education

Australia's colonial mathematics education heritage was not equal in opportunity for working-class people, convicts, girls, or Indigenous students.

> Many groups in society [did] not benefit from having access to any branch of mathematics other than elementary arithmetic. Such groups include females, working-class children, and [others] whose cultures differ from the dominant Anglo-Saxon culture. (Ellerton & Clements, 1988, p. 387)

Male privilege

From the outset, the teaching of mathematics was biased towards young men.

> . . . in the single-sex private and church institutions, the boys' curriculum often included Euclid, algebra, and arithmetic, but in the schools for 'young ladies', arithmetic was the only branch of mathematics studied . . . (Ellerton & Clements, 1988, p. 387)

Schools taught advanced subjects for boys only (NSW Education Department, 1980) because girls were unable to matriculate so there was little point in teaching them more than basic arithmetic. These differences in expectations of males and females remained well into the twentieth century[3].

Class privilege

The NSW Education Department (1980), in its own history of education in early colonial days, noted that:

> [In] the early 1800s, several small academies were begun in Sydney and Parramatta [to cater for boys of] well-to-do parents until [they] were old enough to be sent "back home" to an English boarding school.

These trends continued, with arithmetic being the limit of the mathematical curriculum available for the children of common people.

> . . . by 1855 school mathematics in Australia was well on its way to becoming both a sexist and an elitist affair. Already the tradition that most groups in society would not benefit from having access to any branch other than elementary arithmetic was in place, and a top-down view of the function of mathematics—as a discipline owned by textbook writers, examiners and teachers—existed. (Clements, Grimison, & Ellerton, 1989, p. 57)

In comparison with today's primary mathematics curriculum, though, expectations for primary arithmetic at that time were relatively high. For example, NSW Grade 6 children were expected to be able to multiply numbers with recurring decimals such as $247.\overline{89} \times 257.\overline{76}$ (NSW Department of Education, 1980, p. 23). However, teaching methods for all subjects were described as:

> [T]he teacher, having set the pupils to work, hears each one in turn recite the lessons without the aid of the textbook. Repetition and memorization played a dominant role in the learning process. (NSW Department of Education, 1980, p. 32)

---

[3] When girls were accepted into English universities in 1867, Australian universities followed.

"White" privilege

English was a second or further language for Aborigines in the eighteenth and nineteenth centuries, but was the only language of instruction. Most schools banned children from speaking their native languages. As mathematical ideas are held, expressed, and shared via language, foreign mathematical concepts were being learned in a language unknown to Aboriginal students[4].

There were some attempts to establish schools for Indigenous children only. Opened in 1814, the "Native Institution" at Parramatta was the first (Commonwealth of Australia, 1997, p. 33). The report of the *National Inquiry into the Separation of Aboriginal and Torres Strait Islander Children from their Families* claimed that it aimed to:

> provide them with the 'benefit' of a European 'education' and inculcate the diligent subservience thought desirable in servants and the working class. It was quickly boycotted by Indigenous families. By 1820 it had closed and other attempts were similarly short-lived. (Commonwealth of Australia, 1997, p. 33)

Many students were brought to the school by force (Brook & Cohen, 1991). The school's curriculum would have seemed quite irrelevant to young Indigenous people (Christie, 1979), and "within a few years it evoked a hostile response when it became apparent that its purpose was to distance the children from their families and communities" (Commonwealth of Australia, 1997, p. 23)[5].

A further experiment was at "Liverpool Orphan School", a boarding school in Sydney that included "orphans" removed from their families along with poor non-Aboriginal children. Children surrendered to any school, orphanage, or institution would remain under its authority until the age of 21 regardless of parents' wishes and managers could apprentice out children over the age of 12 (Commonwealth of Australia, 1997, p. 85).

---

[4] In 1788, there were approximately 700 Indigenous languages (Australian Museum, http://australianmuseum.net.au/indigenous-australia-introduction.)

[5] Later, the school was re-established successfully with Maori, Aboriginal, and non-Aboriginal children attending racially segregated classes (Brook & Cohen, 1991).

Aboriginal parents hid their children and students frequently ran away from the school.

Indigenous parents were blamed consistently for neglecting children, and schools were used as a supposed remedy for this.

> Simply being Aboriginal was proof of neglect and for the purposes of the Act missions were declared to be industrial schools or reformatories to which Indigenous children could be sent. . . . The children's attendance was procured in various ways—some came into the School because they preferred a regular supply of food . . . especially in the winter season; others were sent by the parents on condition that they receive a blanket for three months' attendance and others again were sent by the police if found begging about town. (Commonwealth of Australia, 1997, p. 62, p. 103)

In the 1840s, some church-run boarding schools for Aboriginal children were opened with government financial assistance, but by 1847 all but one had closed. In summary, the result was "a cross generational pattern of alienation from schools" (Commonwealth of Australia, 1997, p. 485).

There were several subsequent attempts to establish missionary schools, but these too were unsuccessful and in 1837 a Select Committee of House of Commons (England) inquiring into "conditions of Indigenous people in British Colonies" resulted in appointment of "Protectors for Aboriginal people". However, in 1849 a NSW Select Committee decided against investing in Aboriginal education on the basis that it was "futile" (Fletcher, 1989, p. 34).

Clements, Grimison, and Ellerton (1989) report that the only mathematics taught in schools that accepted Indigenous students echoed that of schools for the poor in England: essentially basic addition, subtraction, multiplication, and division. Bishop (1990) noted that mathematics in colonial times was:

> . . . one of the most powerful weapons in the imposition of western culture. . . . From colonialism through to neo-colonialism, the cultural imperialism of western mathematics has yet to be fully realised and understood (p. 59).

Schools that did admit Aboriginal students usually practised strict physical segregation in classrooms but Aboriginal children could not attend schools if the local "whites" protested[6]. It was not until the 1975 *Racial Discrimination Act* that recognised Aboriginal people as equal under Australian law that full rights to education were endowed. Even today, Indigenous people face prejudicial attitudes in educational contexts. In fact, Grant (2016) claimed recently that, "An Indigenous child is more likely to be locked up in prison than they are to finish high school".

Colonial developments

Other penal colonies were established independently or carved out of NSW, and each started schools. First, "Hobart Town[7]" was established as a penal settlement in 1803 and a *Board of Education* was formed to introduce non-denominational schools[8].

A further penal colony was established near what is now Brisbane in 1824 and a primary school taught by a soldier's wife opened in 1826. The area was soon opened to free settlers, which resulted in an increased demand for education. In 1859, the colony of Queensland was formed, with its own parliament, and a *Board of National Education* was established to create public schools modelled on the "National Schools" of Ireland. This new colony pioneered a public secondary education system in the early 1860s when the government set up "grammar schools" open to both boys and girls of all classes, the first free secondary education in Australia, but this did not include Aboriginal children.

England claimed the Swan River Colony, now Perth in Western Australia, in 1828 but there were just a few primary schools before the gold rushes in the 1850s and older children travelled by ship to Sydney or England for education.

The land that now forms the state of South Australia was proclaimed as a convict-free colony in 1834 and its first school was opened in 1836. Its

---

[6] In fact, it was not until 1967 that Aboriginal people were even recognised as Australian citizens.
[7] Originally called "New Norfolk" and now Hobart in the state of Tasmania.
[8] Tasmania became the first colony to introduce compulsory primary education in 1868, but Indigenous children were not included.

Education Board was established in 1856, but a mass exodus of half of the population to the Victorian gold fields did not help the scarcity of teachers.

Meanwhile, in 1851 the southern area of New South Wales initially known as Port Phillip had become the Colony of Victoria, with its own government. Its constitution of 1855 provided for government funding for public and religious schools. A *Board of Education* was established in 1862[9].

Each of the colonies controlled its own curriculum, but they all copied English content (Grimison, 1995).

Even though it was not unusual for denominational schools to be established in towns as a reaction to the proposed opening of a government school (NSW Department of Education, 1980), initially there had been little argument between systems because all educators followed English traditions and the whole colony of NSW was under the Governor's control[10].

> The economic and social arguments for establishing schools [that] all children in an area could attend ran headlong into equally valid arguments that separate denominational schools were required to teach the particular doctrines of the various denominations. (NSW Department of Education, 1980, p. 12)

However, there were increasing arguments over funding, curriculum content, the role of religious instruction, and responsibility for standards between *five* rival educational systems (the public system plus four denominational ones). State versus church arguments over funding and the role of religious education culminated in the 1830s but continued throughout the next half century[11].

Anglican schools were temporarily privileged:

> The link between the State and the Anglican Church became so strongly forged that in the 1820s a corporation was formed by royal letters patent and given one-seventh of all New South Wales land for the maintenance of the Anglican Church and 'the education of

---

[9] The Victoria Education Department started in 1873.
[10] Church of England clergy supervised public schools.
[11] These still remain points of controversy today.

the Youth in New South Wales'. Other Christian denominations in the colony fiercely challenged the notion of an Anglican religious and educational monopoly. Their hostility, together with political changes in the mother country which allowed greater freedom for Catholics, and growing objections to the alienation of large tracts of colonial land with no possibility of their development led to the dissolution of the corporation. (Board, n.d.)

In the 1840s, economic depression brought drastic cuts to denominational schools, to the extent that only half the colony's children were under instruction (NSW Department of Education, 1980), and arguments intensified. In the Denominational Schools Board, clerical authorities competed not only with the National Board but also with each other.

> By 1854 it was obvious that the main feature of the dual system was the rivalry between the two Boards, which were, in fact, competing for public favour and patronage. (Relton, 1962, p. 138)

Criticism of denominational schools by a select committee of inquiry did not help matters:

> With more trained teachers, the National schools could obviously provide a better type of education, and nowhere was this more manifest than in the religious instruction given in the respective schools—the National schools' "general" religious education seemed notably more suitable than the churches' sectarian education, which would only intensify religious antagonism (Relton, 1962, p. 140).

In 1848, there was an attempt in NSW to settle conflicting interests of "denominational education against secular education, ecclesiastical control against State control, sectarianism against unity, tradition against liberalism" (Relton, 1962, p. 133). The dual system was formalised, with a *Denominational Schools Board* overseeing religious schools and a *Board of National Education* managing secular education. Because of the gold rushes, government subsidies had increased, and that also helped ease tensions. The Public Schools Act of 1866 replaced the two boards with one *Council of Education*. Denominational schools retained some government funding but were to follow the same course of instruction as the public schools and to be subject to the same regulation[12].

A renewed economy and the growth of secondary schools

The gold rushes saw huge growth in Australia's population through immigration, as well as significant progress in economic development. By 1857, NSW had 62 government schools and by 1872 most of the colony's 90 000 non-Aboriginal children were attending school, although regular attendance was a problem, especially in rural areas where potential work included seeking gold and agricultural labouring[13].

In the 1880s, about half the Colony's children were in government schools, one third of the remainder were in Catholic schools, and the rest were shared between other denominational schools[14]. England had expanded secondary schooling to cater for growing industrialisation, so there was demand by parents for the colonies to do likewise (Board, n.d.). Evening public schools tried catering for adults and adolescents without primary education, but classes were short-lived. However, public "technical education" (secondary schooling with an emphasis on trades) commenced successfully. Public secondary schools were established in 1883 to fill the gap between primary education and technical schools or professional apprenticeships[15]. In government secondary schools, students were usually divided into two groups—preparing for the Junior Public examinations and preparing for the Senior Public examinations.

Some "superior public schools" were nominated in Sydney (four pairs for girls and boys) and Melbourne (two schools for boys), and these were the first public schools to challenge higher-level curricula of

---

[12] This is the situation in Australia's education system today.

[13] In 1880, it was regulated in the colonies of NSW and the Port Phillip district that children must attend school for 140 days per year, although this did not apply to Aboriginal children.

[14] Currently, about 70% of Australian children attend government schools, about 18% are in Catholic schools, and the remainders attend independent (private/charter) schools. Independent schools—both religious and secular—charge fees; but all are now subsidised by both Federal and State governments.

[15] Public high schools, usually more selective but cheaper than denominational schools, appealed to the new middle classes.

denominational schools[16]. However, the establishment of these and
continuation schools ("higher departments" of existing state schools) that
included the study of mathematics other than arithmetic challenged class
privileges, so was not a simple step. A biographer of an influential
educational administrator in Victoria, Frank Tate (1864–1939), detailed
the following problems.

> In fact it immediately placed [public schools] in competition with the
> private secondary schools, offering the same or a wider range of
> subjects and publicizing its matriculation results with similar pride.
> Defenders of the private schools . . . denounced this state intrusion
> into education as 'simply Socialism'. . . . Tate gradually [won] the
> argument that the state should enter secondary education to ensure
> that it was equally available to all children. (Selleck, 1990, n.p.)

Tertiary developments

Although university science subjects were available to students before
1870, there were no whole science degrees. The 1880s saw developments
in the universities in Sydney, Melbourne and Adelaide, including
commitments to science degrees and the consequent appointment of
science professors, leading to "a trickle of research students in the physico-
mathematical sciences" (MacLeod, 1998, p. 151). However, school
teaching, surveying, and astronomy aside, there were few jobs for
physicists or mathematicians so few people chose mathematics as a
university subject.

The colonial legacy

To date, Australian colonisation had meant that:
1. Few sections of the population studied any branch of mathematics
   other than arithmetic, so school mathematics became elitist, sexist,
   and racist;

---

[16] Mathematics in these schools included arithmetic, algebra, bookkeeping, mensuration,
and trigonometry—but girls did not study it.

2. Written examinations, with payment-by-result for teachers, dominated curriculum development, teaching approaches, and assessment;
3. Few primary teachers went beyond Grade 8 in their own education before undertaking apprenticeships as junior teachers; and
4. A Euclidean approach to geometry dominated university entrance exams and hence secondary teaching.

(Paraphrased from Ellerton & Clements, 1988)

After the federation of the colonies to form the Commonwealth of Australia in 1901, they became self-governing states with separate Education Departments[17]. Generally, department officers set curricula for primary schools while state public examination boards determined secondary content. For mathematics, these boards were advised by syllabus committees made up of secondary and tertiary representatives, usually chaired by the relevant capital city university's Professor of Mathematics (Blakers, 1978). Thus each state maintained control over education, and has done so through to modern times[18].

There was not full agreement across the states on secondary mathematics curricula, nor adequate teacher training. Further, Australian mathematics teaching was constrained by outdated resources. For example, non-Euclidian geometry had been accepted into the English secondary curriculum as well as university entrance exams early in the 1900s, but texts used in NSW in the first half of the twentieth century had been published in England up to fifty years prior (Grimison, 1982).

> [The] typical [mathematics] offerings . . . included predominantly Euclidean geometry, algebra, and some arithmetic, and a little mechanics, conics and elementary trigonometry. . . . A number of "tricks" were incorporated in the anticipation that students would memorise the procedures and hence be able to reproduce the method in answering an examination question. . . . Most texts

---

[17] Territories were each managed by an appointed state.
[18] A federal Department of Education achieved increasing control during and after the twentieth century.

included questions from previously set [English] examinations. (Grimison, 1995, p. 321).

Teachers were paid by results, but by 1910 (NSW) and 1914 (Vic) district-based inspectors were appointed to monitor teaching quality in public schools. Independent schools could elect to participate in the inspection scheme, with those that did not being classified as "B schedule". (Schools on the A schedule could pass candidates for Intermediate and Leaving certificates [Blake, 1973].)

The University of Sydney gradually vested control of examinations (and hence of secondary curricula) to new universities. For example, public examinations in Western Australia dated from 1895 when its independent schools persuaded the Universities of Sydney and Adelaide to open an examination centre in Perth (White, 1975). Likewise, the University of Queensland, opened in 1911, was given power to conduct public examinations and issue certificates.

Most professions (including primary teaching and the Public Service) required two—later three, then four—years of post-primary education[19], while university entrance required five[20] (the "Leaving certificate").

While curricular and examination content were the same for all students:

> The transformative period of "secondary schooling for all" did not mean equality of educational opportunities through extended access to secondary education. . . . [Completion] rates for secondary schooling were 25% lower for students attending schools in the working class suburbs than for those attending schools in middle class locations. Moreover, for both locations, girls had completion rates that were half those of boys, suggesting that girls were still being 'ghettoed' into terminal streams in high schools, and dropping out of school earlier and more frequently than boys. (Australian Bureau of Statistics, 2001, n.p.)

---

[19] "Intermediate certificate".
[20] The five years for matriculation became six years during the 1950s (Vic.) and 1960s (NSW).

Figure 1 summarises the types of secondary schools in NSW in 1913[21].

There was a huge, but not steady, growth in post-primary enrolments in the first half of the twentieth century in all states. Figure 2 shows public post-primary enrolment growth from NSW (only). The significant dip early in the 1930s was a product of the "great depression", when many teenagers took any paid work available.

In 1944 in Victoria and 1962 in NSW, secondary schooling was increased to six years, with the last two years being called "Higher School Certificate". Universities' Schools Boards maintained control over matriculation examination content and pre-requisites for university entrance, which had a significant impact on the content for Leaving and Intermediate exams as well as preparatory years.

The federal government began to offer scholarships for tertiary students in 1951 and secondary students in 1956. It also accepted more financial responsibility for funding universities about this time, and federal aid was expanded in independent primary and secondary schools via per-pupil funding. The Commonwealth's role in curriculum control, however, remained a "sensitive political issue" (NSW Department of Education, 1980, p. 214).

Teacher training

In the earliest colonial times schools used untrained teachers, but as immigration to Australia increased some trained teachers from England and Scotland took up teaching positions in denominational schools (Partridge, 1968). By 1850, teacher training had become a government priority, and "model schools" were established in NSW and Victoria (Hyams, 1979)[22]. One model school with trainee teachers had 800 secondary students plus 600 correspondence students and 400–600 students in evening classes, thus "influencing the lives of 2000 young persons every week of the year" (Blake, 1973, p. 777).

---

[21]The 14 subjects available for study at Leaving (Year 11) level in 1912 included Algebra (studied by 52% of students), Geometry (42%), and Trigonometry (46%).

[22] A "normal school" was later developed for teacher training in Brisbane (1862) and Tasmania (1865). Professors gave lectures to teachers on Saturday mornings.

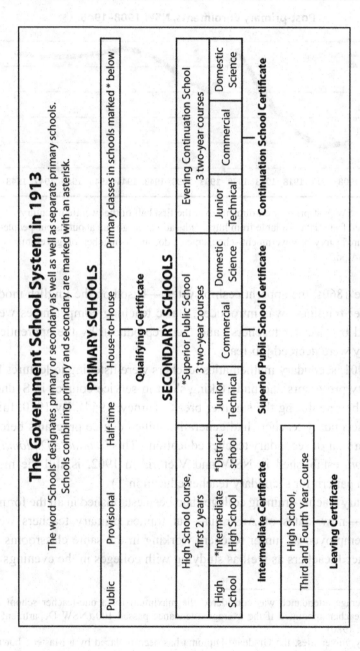

Figure 1. Types of schools run by the NSW government in 1913 (from Board, n.d.).

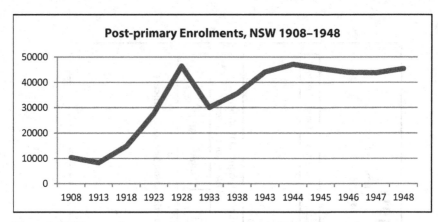

Figure 2. NSW post-primary enrolments over the first half of the twentieth century (constructed from data available from https://education.nsw.gov.au/about-us/our-people-our-structure/history-of-government-schools/media/documents/number_of_government_schools_AA.pdf).

In the 1860s, the apprenticeship system—based on the English model of teacher training—was introduced. Basic teaching competencies were modelled by "Master teachers" and then reproduced by the apprentices until they were decreed job-ready[23].

In 1901, secondary mathematics teachers were listening to lectures by university professors and undertaking short in-service courses in Sydney and Melbourne during the summer break (Turney, 1964). By 1910, fully trained secondary teachers had generally studied degree programs before one extra year of secondary teacher education. (This *Graduate Diploma of Education*, established in NSW and Victoria in 1902, is still the most common pattern for secondary teacher education[24].)

Primary teacher-training colleges had been established in all the former colonies—now states of Australia—but trainee primary teachers were usually employed as junior teachers working in the same classrooms as experienced teachers as well as studying with colleges in the evenings or

---

[23] 100 average attendance was considered the maximum for a one-teacher school, an assistant teacher appointed if the average attendance passed 100 (NSW Department of Education, 1980).
[24] In many universities, the Graduate Diploma has been replaced by a Masters, but the model is still 3+1 or 3+2 years.

by correspondence. The elite university system remained disconnected from primary teacher education that was embedded in the training colleges that ranked as second tier institutions.

Professional associations for teachers of mathematics

In England in 1870, an association was formed to study and reform the teaching of mathematics (Veness, 1985). By 1897, this *Association for the Improvement of Geometrical Teaching* had broadened its interests and became *The Mathematical Association* (Veness, 1985).

Independently, a *Mathematical Society* was founded in Melbourne in 1906 by a group of academics, using university lecture rooms with notices of meetings being published in Melbourne papers. After five years, the society's minutes became sketchy until the group was revived again in 1911 as the *Mathematical Association of Victoria* [MAV], an affiliate of England's *Mathematical Association*. From 1911, MAV's minutes show growing concern by the academics about the content of the Leaving Certificate curricula. In 1926, MAV also made recommendations about early secondary mathematics (i.e., less symbolism in early algebra and no teaching of statistics in schools) as well as about methods of teaching: "Every lesson should start with some mental work and that work should be corrected largely by the pupils" (MAV minute, October 1926). In time, there were also general recommendations about primary arithmetic and the preparation of teachers for this[25].

There were nine branches of *The Mathematical Association* in Australia by 1911, including the forerunner in Sydney of the *Mathematical Association of New South Wales* [MANSW, founded 1910] with a journal entitled *The Australian Mathematics Teacher* [AMT]. There were several proposals to incorporate the various state associations into one, but it was not until 1966 that an umbrella body, the Australian Association of Mathematics Teachers [AAMT[26]] was formed and took over editorial

---

[25] The MAV is still both active and influential in directions for primary and secondary mathematics teaching.
[26] Individual people and schools do not join AAMT: they join the relevant state association and hence gain access to AAMT benefits and services. AAMT is a federation of autonomous bodies.

control of *The Australian Mathematics Teacher*—Australia's first national journal for teachers.

Today, the national association and its state-based affiliates are active in the provision of resources for teachers, professional development, special days on aspects of tertiary mathematics aimed at talented students, competitions such as the *Mathematics Olympiad*[27], newsletters and bulletins, comprehensive web-based resources, syllabus and examination discussion days, and annual conferences for all teachers of mathematics. The AAMT and the *Mathematics Education Research Group of Australasia*[28] [MERGA], provide regular opportunities for interstate co-operation, professional meetings, research, and conferences. *The Australian Mathematical Sciences Institute* [AMSI] was established in 2002 in response to a need for national collaboration across the mathematical sciences.

Many current Australian mathematics educators are also active members of international associations such as the *International Group for the Psychology of Mathematics Education* [PME] and the *International Commission on Mathematical Instruction* [ICMI], making significant contributions to research and leadership worldwide.

Teacher shortages

After World War 2, the growth of Australia's population necessitated more teachers' colleges. (By 1964 there were 29 teachers' colleges throughout Australia compared with nine before World War 2 [Hyams, 1980]).

National increases of about 90 000 pupils per year in the 1950s led to large classes. The introduction of equal pay for female teachers attracted more teachers, standards of entry to training colleges were lowered, the duration of courses was shortened, the apprenticeship model was reintroduced (Dyson, 2005), and untrained mathematics teachers were employed in many secondary schools. Many new primary and secondary Catholic schools were opened, employing lay teachers. Migrant secondary

---

[27] The Australian Mathematics Competition is one of the largest student competitions in the world. Entrants come from more than 30 countries.
[28] The *Mathematics Education Lecturers Association* [MELA] was subsumed by MERGA in 1996.

teachers with expertise in discipline areas—including mathematics—were imported. Regional education offices were established to cope with Department workloads, and inspection of teachers was phased out. In some states, school councils (which included teachers and parents) were given more control over the curriculum, school maintenance, and minor financial matters. As the peak of the post-war "baby boom" worked its way through primary then secondary schools, demountable classrooms were added to most schools.

Teacher registration

In 1974, the Queensland government funded the establishment of an inter-systemic *Teacher Registration and Accreditation Authority*—the first of its type in Australia[29]. State registration boards soon had an impact on the content of teacher education courses through setting requirements for registration of graduates, although given the independence of universities this was a contentious point.

While states initially maintained the responsibility for their own teachers' registration, the federal government recently became involved in initial teacher education with the formation of the *Australian Institute for Teaching and School Leadership* in 2010, with state Ministers of Education agreeing on standards and procedures for teacher education course accreditation (AITSL, 2011).

Professional development

Both state and federal governments support short professional development courses[30]. Teachers need to apply for re-registration periodically, requiring completion of a number of professional development courses every few years, so the demand is high[31]. Universities continued to offer further degrees to teachers.

---

[29] It was thirty years before all other states had teacher registration boards.
[30] In the 1970s, professional development became relatively well funded.
[31] Most mathematics professional development courses are offered by universities or professional associations, and some are on-line so are accessible to teachers in remote areas.

Post-compulsory education and training are now regulated within the *Australian Qualifications Framework* [AQF], a unified system of national qualifications in the vocational education and training sector and universities, so differences between institutions are diminishing.

Long before this time, though, universities had made their own progress.

Some Key Changes During and Since the Twentieth Century

In their summaries of mathematics education in the nineteenth and twentieth centuries, Ellerton and Clements (1988) noted the ubiquity of "rote teaching and learning procedures associated with rigidly defined courses of study, prescribed text books, and written examinations" (p. 387). However, there have been some major changes in both primary and secondary mathematics curricula and teaching during the second half of the twentieth century and subsequently. These included greater use of problem solving and modelling, increased attention to girls' success in mathematics classrooms, and the "New Maths" movement.

Problem solving and modelling

The mid-twentieth century saw challenges to "pure" mathematics, with more emphasis on applications, real-world problem solving, and mathematical modelling.

Problem solving became a "buzzword" in Australian primary and secondary curriculum documents, classrooms, and teacher education as well as in the publication of resources for teachers. This movement came from (a) the UK, where the *Mathematics Applicable* (e.g., Ormell, 1971) group was influential and the "Cockroft Report" recommended that "The development of general strategies directed towards problem solving and investigations can start during the primary years" (Cockroft et al., 1982, p. 95); (b) the USA, where The *National Council of Teachers of Mathematics* (NCTM) had recommended that "problem solving be the focus of school mathematics in the 1980s" (National Council of Teachers of Mathematics, 1980, p. 1); and Europe where the Netherlands' *Realistic Mathematics* was influential.

For example, problem solving became *the* key change in Year 11–12 mathematics in Victoria as a result of the *Ministerial Review of Postcompulsory Schooling* (Blackburn, 1985). The review report included 45 recommendations that were implemented, the major one being that a single two-year certificate marking the completion of secondary schooling named the *Victorian Certificate of Education* [VCE] be introduced. Here, the aims were full participation and retention of students with the inclusion of practical studies for a broad education. Mathematics subjects were to be examined in both years 11 and 12, but Blackburn argued for "a reconsideration of mathematics subjects in contemporary terms" (p. 22) with much emphasis in exams on solving realistic problems. This movement resulted in much more emphasis on problem solving in the lower secondary years nationally but did not seem to result in significant longer-term changes to pedagogy or assessment.

Twentieth century primary arithmetic teaching approaches had already been influenced by the use of manipulatives to model problems—especially by using Montessori, Cuisenaire, and Dienes' materials. Each of these theorists was temporarily influential, particularly in the 1960s and 1970s when there was emphasis across Australia on the manipulation of concrete materials. Heavy use of a wider range of manipulatives has continued in Australian primary mathematics education until today, but few were (or are) used in secondary schools where the use of mathematical modelling has more to do with a growth of technology use.

"New maths"

For both primary and secondary teachers and students, a further import of ideas and materials was the 'new maths' movement of the early 1960s, putting emphasis on algebraic structure and inequalities, properties of operations, Boolean logic, deduction, functions, transformation geometry, and set theory. Its implementation in Australia was decided on during the 1964 conference of the *Australian Curriculum Officers of State Education Departments*.

New mathematics courses were developed in Australia in the early 1960s, and a rash of new textbooks, written by Australian authors and containing relatively little formal Euclidean geometry, quickly appeared. These placed great stress on the use of axiomatics in the teaching of number. (Ellerton & Clements, 1988, p. 392)

This movement challenged most primary and secondary teachers, teacher education students and pupils, and lack of support from most universities saw it die out within a decade. It soon became fashionable for teachers and parents alike to reject "new maths"[32], although the professional associations had benefitted from opportunities to support teachers with professional development courses (Veness, 1985), the first formal conferences, and research. Further, despite the *Australian Council for Educational Research* [ACER] having been established in 1930, it seems that it conducted little substantive mathematics education research in Australia before "new maths".

Girls' participation

The publication of *Girls, School and Society* (Schools Commission, 1975) was a milestone in acknowledging that gender differences needed redressing. Mathematics teachers became more aware of girls' needs and encouraged girls to choose mathematics, or harder mathematics subjects, in Years 10 to 12.

Each state used its own strategies to encourage better choices; e.g., Victoria showed advertisements portraying girls trapped into boxes, with the voice-over explaining girls were limiting their options by not choosing mathematics.

Towards the end of the twentieth century, Barnes (1991) addressed the problem of girls not choosing calculus through the provision of a text that explained it in terms that were "user friendly". The use of group work, meaningful examples, and discovery learning were also portrayed as vital. One teacher-researcher who used Barnes' text in his Year 11 classroom reported that:

The students achieved a greater appreciation of the historical development of calculus and the ways . . . it can be used to solve

---

[32] A back-lash back-to-basics movement resulted.

everyday problems. They also reported positive feelings toward their work and generally found it more interesting and enjoyable than other topics they had studied (Cavanagh, 1996, p. 113).

The enrolment in Year 12 of males and females in mathematics units from 1980 to 2004 is shown in Figure 3. The 2004 data showed a higher percentage of female than male students continuing into Year 12, so the situation seemed promising.

However, overall enrolments in the higher-level mathematics subjects decreased and there continued to be increasing percentages of males in higher-level mathematics (see Figure 4) and increasing participation by females in lower-level mathematics.

The downturn that Forgasz noted was due partly to lower percentages of students completing school to Year 12. After a period of dramatic growth in school completion in Australia to 1992 (77%), rates of school retention had begun to decline. The reasons included a global recession with consequent reductions in labour market opportunities and less demand for higher education places. Lamb (1998) noted that this decline affected some social groups more than others and proposed that:

> Larger class sizes, general staff cuts, fewer resources, reduction in support services, and the loss of specialist teachers may well have had an impact on the range and quality of school programs and, ultimately, on the quality of teaching and learning experiences of students (Lamb, 1998, p. 48).

More flexible tertiary admission

As noted above, university entrance requirements—particularly of the more prestigious and influential "sandstone" universities in capital cities—influenced secondary-school mathematics curricula. In the 1970 to 1990 period, newer universities challenged this dominance by providing alternative pathways and "bridging" units that allowed mature-age students without higher-level mathematics units to bridge the gap into courses that use tertiary-level mathematics.

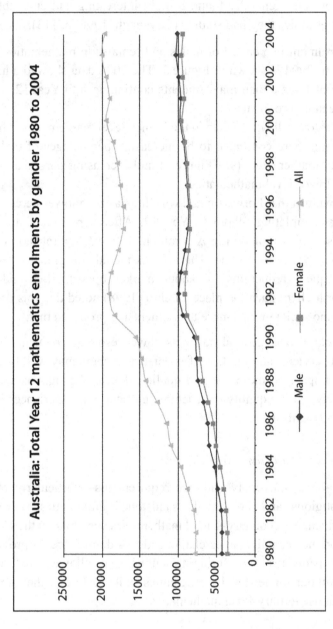

Figure 3. Australia: Total Year 12 mathematics enrolments by gender 1980 to 2004 (from Forgasz, 2006, p. 20).

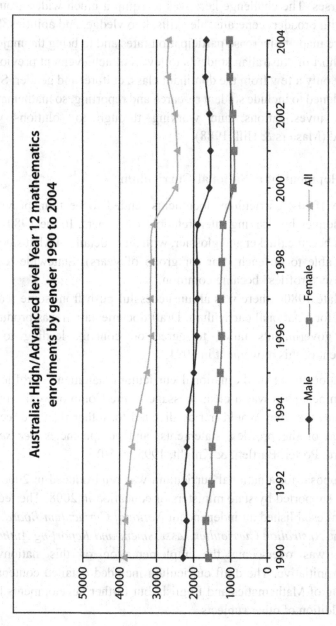

Figure 4. Australia: High/Advanced level Year 12 mathematics enrolments by gender 1990 to 2004 (from Forgasz, 2006, p. 17).

This period also saw a plethora of state reviews of the content of Year 11–12 courses. The challenge here was to equip a much wider group of students with broader, generalisable skills, knowledge, and abilities. New efforts were made to increase participation rates and to bring the majority of each cohort of Australian students to levels of achievement previously reached by only a few from the dominant class, culture, and gender. Skills were broadened to include student research and reporting, so mathematical modelling, investigations, and working through to solutions were emphasised (Masters & Hill, 1988).

The Development of a National Curriculum

Before the 1980s, curriculum documents tended to be thin, providing broad guidelines but leaving interpretation to teachers. In the 1980s, the documents became thicker and glossier, with more details of what students should be able to do each year (or group of years). Statewide testing against these "profiles" became common.

In the late 1980s, there was an unsuccessful push from some federal politicians for a national curriculum. Draft documentation was produced, but state governments failed to agree on content, leading to the abandonment of this movement in 1991.

> Non-endorsement of the national curriculum statements, profiles and competencies was a clear message to the Commonwealth by the States that they would not be dictated to, rather than a direct rejection of the profiles, statements, and competencies *per se.* (Lingard, Porter, Bartlett, & Knight, 1995, p. 50)

The proposal for a national curriculum was reintroduced in 2006 and eventually supported by state ministers of education in 2008. The federal government established an independent *National Curriculum Board*, and by 2009 the *Australian Curriculum, Assessment and Reporting Authority* (ACARA) was overseeing the implementation of this nationwide curriculum initiative. The draft curriculum included detailed content for the teaching of Mathematics and English, but further developments have seen the addition of other subjects.

The Commonwealth has claimed a lot of control over K–10 content through its use of *National Assessment Program—Literacy and Numeracy* [NAPLAN] (Australian Curriculum, Assessment and Reporting Authority, 2014) where there is national testing of students in Years 3, 5, and 9[33]. NAPLAN results for individual schools are now published, with other school information, on the federal "My school" website[34]. NAPLAN results are also publicly reported through summary and progressive result reports that are available for use by jurisdictions, non-government school sectors, and schools. For parents:

> Individual student reports . . . show student results against the national average and the middle 60 per cent of students nationally. These reports contain a description of what was assessed in each of the tests and provide information about the knowledge and skills the student demonstrated in the tests (ACARA, 2014, p. iv).

The *Australian Curriculum (Foundation – Year 10)* experienced compatibility issues with some existing state curricula, and states did not wish to lose responsibility for curriculum development, so each state has since developed a range of curriculum guidelines and support documents for teachers. Each has also retained its Board of Studies or equivalent to control Year 11–12 content and standards. Most independent schools follow national curriculum guidelines and most participate in NAPLAN testing, but some have chosen to offer the *International Baccalaureate* (IB).

Mathematics Teaching in Australian Schools Today

The various topics in Australian school mathematics (arithmetic, algebra, geometry, etc.) have generally been taught in both primary and secondary schools concurrently as "Mathematics"[35], at least since the mid-nineteenth century. Trigonometry and calculus are generally added in the middle- and upper-secondary years respectively for stronger students. Probability and statistics were taught in some states, but introduced nationwide at both primary and secondary levels in the National Curriculum.

---

[33] A set of sample NAPLAN tests is available from https://www.nap.edu.au/naplan/the-tests.
[34] https://www.myschool.edu.au/
[35] The term "Numeracy" was a temporary fad in some states.

Across Australia, students choose which subjects to study in Year 11 and 12, and some decide not to study any Mathematics at this level. Information about the national curriculum for years 11 and 12 is available from   http://www.australiancurriculum.edu.au/seniorsecondary/overview (see "Mathematics pdf" link) but state and territory authorities still "determine assessment and certification specifications for their courses and any additional information, guidelines, and rules to satisfy local requirements" (ACARA website).

On-going positive developments

Clements, Grimison, and Ellerton (1989) point out ways that Australian mathematics education has now thrown off its colonial shackles.

> . . . certainly we could refer to (a) the high level of cooperation between mathematics educators in different states and territories, (b) the high quality and distinctive character of recent Australian mathematics education projects, (c) the improved preservice and inservice professional development of teachers, (d) the greater awareness of, and planning for, cultural factors (including language backgrounds) which influence mathematical learning, (e) the development of mathematics curriculum frameworks, (f) moves for increasing the access and success of previously disadvantaged groups . . . , and (g) the development of new assessment procedures (Clements, Grimison, & Ellerton, 1989, p. 71).

There have been some identifiably positive changes since the 1990s, outlined below.

The impact of electronic technologies

Recently there has been quite a bit of research into the use of computer algebra systems in Year 11 and 12 curriculum and assessment. Garner (2016) noted that "Mathematical Methods CAS Units 1–4" were introduced in the state of Victoria as part of the *Victorian Curriculum and Assessment Authority [VCAA] CAS Pilot Study* and that "each of the classes involved has undergone a clearly observable change in the way teaching and learning takes place" (Garner, 2005, p. 120). The need to make use of digital technologies in learning mathematics is now included in most state curriculum documents within Australia (Goos, 2010), suggesting an urgent need for relevant professional development. This applies to training in the use of calculators in primary schools as much as it does for the use of scientific then graphing calculators and computer software in secondary schools, because pedagogy at one level impacts on other levels. In the senior secondary years, the current national curriculum documents allow appropriate use of computer algebra systems and dynamic geometry.

Whether to allow the use of computer software use in senior-secondary pedagogy and assessment varies, though, according to state. Queensland allows any mathematics technologies in assessment. Western Australia and Victoria provide with-technology options for subject choice in Years 11 and 12, with both calculator-free and calculator-assumed exams. South Australia allows only graphics calculators, and New South Wales allows only scientific calculators in exams.

Developments in Indigenous education

Given the history of education being imposed on Indigenous peoples with an assimilationist agenda, the current movement in education is for educational institutions, at all levels, to build positive relationships with Indigenous people and incorporate Indigenous knowledges and perspectives into aspects of the curriculum. The purposes are to ensure that a) Indigenous Australians are represented, have a voice, and are valued in the curriculum; and b) non-Indigenous Australians gain knowledge about and value Indigenous peoples and cultures.

ACARA included a "cross curriculum priority" within the national curriculum, "Aboriginal and Torres Strait Islander Histories and Cultures" (ACARA, 1916). This means Indigenous-specific content is to be included in all areas of the curriculum. However, the "priorities" are not complusory and, to date, very little information and/or exemplars have been provided for mathematics.

There are two approaches that are currently being implemented in some schools. The first is *Stronger Smarter,* developed by Dr Chris Sarra from his experience as the first Aboriginal principal at Cherbourg State School in the Cherbourg Aboriginal Community, Queensland. This approach focuses not directly on curriculum or classroom pedagogy but on high expectations, positive sense of cultural identity, and positive Indigenous leadership. It emphasises whole school change that will build relationships with Indigenous people, develop leadership positions in the school, and from these partnerships create a culture of high expectations and positive identity for students (Sarra, 2011).

A second initiative, the *Cape York Partnership*, was set up by Mr Noel Pearson to tackle a range of issues in Indigenous education. Pearson has played a significant role representing Indigenous issues to parliaments and the nation. His vision is for "a fully bi-cultural education, to enable [students] to move between their home worlds and the wider Australian and global worlds and enjoy the best of both" (Pearson, n.d.). The reform involves use of "Direct Instruction", based on the belief that every student can achieve academically if they receive effective instruction. Students are placed in ability groups and direct instruction involves a "process of repetition, replication and reproduction of received knowledge" (Kalantiz, 2006, p. 17 cited in Ewing, 2011). As suggested by Ewing (2011), Direct Instruction may not support Indigenous students' engagement in mathematics since its didatic teaching style disempowers the learner, which is not in-line with the original vision of the Cape York Partnership.

The *Make It Count* project was the first large-scale, widespread mathematics education project for Indigenous learners across Australia. The project was developed and implemented by AAMT for three years (2009–2011). It focused on 26 urban schools with high proportions of Aboriginal students, with eight clusters of schools across four states. Each cluster had an assigned mathematics education academic who worked with teachers, Aboriginal education workers, and communities to develop new ways of teaching mathematics to Indigenous students. The findings of the *Make It Count* project are available from http://mic.aamt.edu.au/ —for each cluster, as well as an overall synthesis that is intended to be transferrable across different educational contexts. One of the main outcomes was to develop "responsive" mathematics pedagogy that is culturally, academically, and socially inclusive.

Another outcome from *Make It Count* was the establishment of the *Aboriginal and Torres Strait Islander Mathematics Alliance* [ATSIMA]. ATSIMA has the vision that "All Indigenous learners will be successful in mathematics" (ATISMA, n.d.), and it has been involved in a range of activities to raise the profile of mathematics education for Indigenous learners nationally. It supports the development of culturally responsive mathematics pedagogy and holds national conferences.

The latest development in Indigenous education has involved a large investment by BHP Billiton aimed at improving Science, Technology, Engineering and Mathematics [STEM] education for Indigenous leaners. BHP Billiton contracted the Commonwealth Scientific and Industrial Research Organisation [CSIRO] to develop and implement a suite of STEM education projects. One of these projects is *Prime Futures* that started in 2015 and aims to improve mathematics education for Indigenous learners from Foundation to Year 9 in clusters of schools across Australia. Prime futures uses the *YuMi Deadly Maths* approach (see http://ydc.qut.edu.au/about/about-yumi-deadly-maths.jsp), which is an adaptation of the RMAR model[36], developed by Matthews (2015).

Ongoing concerns for Australian mathematics education

Despite the very positive developments outlined above, there remain some ongoing issues that need to be addressed by all state and territory school systems, including independent schools.

Limited mathematics study time

Given that students study a limited number of subjects at matriculation level, one on-going issue is the need to include pure and applied

---

[36] Also known as the Goompi model.

mathematics in the senior secondary-school years. This has been a problem for many years:

> In the rivalry of intensive studies at this level, where a student can keep up with only four or five subjects . . . if two of the chosen subjects are mathematics there is not room for many branches of science. What we need is a changed approach in mathematics, which is still far too often concerned with formulae and procedures to the exclusion of practical application and interpretation. (Hoy, 1957, p. 10)

Teacher preparation

Whenever there is a top-down change in the curriculum, as there was with the introduction of the 2009 national curriculum, there is a period of uncertainty for teachers and pupils alike as well as parental and societal concerns about a lessening of standards, e.g., in 1901 when the NSW government introduced a "New Education" syllabus, with more interesting and practical mathematics and less emphasis on rote learning, schools were criticised for "soft pedagogy" and "loss of standards" (NSW Department of Education, 1980, p. 139). This doubt results from teachers feeling inadequately prepared for changes as well as pressure from parents to have their children experience the same rigor that they faced.

An overcrowded primary curriculum

There are 16 subjects taught in primary schools, grouped into eight key learning areas[37]. It is recommended that primary children average 174.3 hours per year studying mathematics, but there is general agreement that the curriculum feels overcrowded:

> There is no doubt that the issue that has caused the greatest amount of angst is the amount of content teachers are required to teach. This issue did not come as a surprise. The Australian Primary Principals Association (APPA) has consistently articulated this concern, and it was echoed throughout the consultation process . . . [Their]

---

[37] In public schools and most independent schools, primary teachers (except for specialist Art, Library, etc. teachers) generally teach all subjects to one class of children.

Reviewers are convinced that immediate and substantial action is required to address the overcrowding of the primary curriculum. (Donnelly & Wiltshire, 2014, p. 3)

The national curriculum document *is* comparatively detailed and wordy[38]. However, teachers mainly feel rushed in their efforts to "cover" the mathematics curriculum amongst other subjects, and there is pressure to reduce national expectations. Some state departments are responding; e.g., Queensland's 2015 *Curriculum into the Classroom* [C2C] reduced eight areas of mathematics to six while still claiming "to teach, assess and report using the F–10 Australian Curriculum: Mathematics" (https://www.qcaa.qld.edu.au/p-10/aciq/p-10-mathematics).

Shortages of secondary mathematics teachers

An ongoing problem across Australia is the shortage of well-qualified teachers of secondary mathematics, which continues despite specified visa schemes. This is not a new problem, of course: the 1927 minutes of the MAV show it ran summer schools for untrained teachers who were "lacking, in varying degrees, the basic principles of method" (MAV minutes, July 1927).

Most mathematics teachers are qualified as teachers but many teach out of their discipline area. Percentages of "out-of-field" mathematics teachers vary according to who is reporting, but figures quoted by the Australian Government Office of the Chief Scientist (2014) revealed that 40% of year 7 to 10 students were taught by an out-of-field mathematics teacher in 2012[39]. The *Discipline Profile of the Mathematical Sciences 2015*, produced by the Australian Mathematical Sciences Institute [AMSI] reported:

Difficulty in filling vacancies leads to teachers teaching "out-of-field"; retired teachers being hired on short-term contracts; or, in

---

[38] The whole document has over 1700 pages of text and 568 000 words, but its website is generally user-friendly and does include items useful for teachers such as samples of children's work at the expected standards as well as the quite detailed content descriptors.
[39] Wienk (2015) found that teaching out-of-field is less prevalent in independent schools than in public schools.

acute shortages, teachers not fully qualified in subject areas being recruited to teach these subjects (Wienk, 2015, p. 16).

State and territory governments hide their out-of-field figures in various ways; e.g., the ACT includes in the mathematics field teachers of physical education and science. Strategies used by principals to conceal or deal with shortages include reducing the range of mathematics subjects offered or the length of classroom time for the subject, combining classes within subject areas or across subject areas, and combining classes across year levels (Wienk, 2015). In some rural areas, mathematics teachers are shared between more than one secondary school.

The general unpopularity of science and mathematics

While primary teachers are all qualified teachers, they are generally not strong at mathematics and many do not like it (see, for example, research by FitzSimons, Seah, Bishop & Clarkson [2000] and White, Way, Perry, & Southwell, [2005/6].) Despite the efforts of state education departments, professional associations, and university faculties, many Australian pre-service primary teachers have difficulty with mathematical content at about Grade 8 level (Mays, 2005). Crane (1950) investigated teachers' college students' opinions of secondary courses, finding that Mathematics is one of the two most unpopular subjects because it seemed:

> . . . hard and dull, not clearly explained, and . . . pupils could see little of value in it. Most enlightening is the fact that 70% accepted mental discipline as the main reason for teaching Mathematics, a reflection on the syllabus or on the way it is taught. (Shears, 1957, p. 32)

For secondary teachers, many authoritative statements have been made on the failure rates in Australian universities, with mathematics courses being tailored to engineering students and those aiming for higher degrees in mathematics. Overall, the result of this situation is that the universities are not producing enough graduates to satisfy the teaching needs of secondary schools.

Being culturally responsive

While Australian schools are relatively multicultural, containing many migrant students from Asia[40], the Pacific Islands, and New Zealand, as well as Indigenous Australian students, the traditional curriculum has always favored a Western European version of mathematics and mathematical thinking. From Australia's earliest days, being culturally responsive in the mathematics classroom has not been a central focus of the development of mathematics education!

The biggest system failure here remains the performance of Indigenous students. International assessment of age 15 students' mathematical, scientific, and reading literacy—the *Programme for International Student Assessment* (PISA)—shows that the gap between Indigenous and non-Indigenous students has remained the same for the last decade; i.e., Indigenous are approximately two-and-a-half years behind their non-Indigenous peers. Similarly, analysis of the *Trends in International Mathematics and Science Study* [TIMMS] results by Thomson, McKelvie and Murnane (2006) showed that in primary mathematics Indigenous students achieved, on average, 79 score points lower than non-Indigenous counterparts and 38 score points lower than the international mean. This situation continued until 2015.

Further, the Organization for Economic Cooperation and Development [OECD] has expressed concerns about Australian monitoring of students of lower socio-economic standard and Indigenous status[41].

A report on the *Western Australian Aboriginal Child Health Survey* (De Maio et al., 2005) summarises reasons that apply across Australia for these results:

> Poor educational outcomes among Aboriginal peoples have been evident for many decades and are influenced by a number of factors not shared by other Australians [including] the geographical

---

[40] 12% of the Australian population has Asian ancestry. (Bureau of Statistics, reported in *The Sydney Morning Herald,* Jun 22, 2012: *"Land of many cultures, ancestries and faiths".* Chinese, Indian, and Vietnamese Australians are among Sydney's five largest overseas-born communities. Some city schools have students born in more than 60 countries.
[41] The same report also recommended greater monitoring of schools in the non-government sector.

dispersion of the population; minimal use or knowledge of Standard Australian English (which accounts for significant proportions of Aboriginal children who begin school in remote [areas]); and a high degree of chronic health conditions (De Maio et al., 2005, p. 38).

Even though the issues are complex, simple steps have not been made within education systems. For example, there are many Indigenous students who are raised to speak several Indigenous languages but very little English, but Australia has had no bilingual schools using Aboriginal languages sanctioned by state or territory governments.

The city-country divide

Australia is a large continent with about 90% of its population[42] living in its cities[43]. This creates difficulties in remote rural areas that include filling vacant mathematics teaching positions, attracting and retaining fully-qualified teachers, "professional connectedness[44]" (Lyons et al., 2006, p. vi), rural students having few role models of adults using other than basic arithmetic or proceeding to university, parents frequently having relatively low career expectations for their children, students having limited senior courses to choose from, and the typical use of multi-level classes (Lyons et al., 2006). The teachers appointed to very remote schools are inclined to be less experienced and generally less competent than those in metropolitan schools (Lyons et al., 2006).

As in colonial times, many Australian students suffer more than one layer of disadvantage: socio-economic, cultural, and/or geographical.

Conclusion

Despite these ongoing problems, Australian mathematics education has come a long way in the 200+ years since white settlement. There have

---

[42] At the time of writing, Australia's population is about 24 million.

[43] About 90% of Australia's 7 686 850 km² area is deemed "uninhabitable", although some of this land is used for farming.

[44] Secondary mathematics teachers in remote areas are often the only one in the school or district. Professional development activities usually involve the expense and inconvenience of travel and accommodation.

been many distinctly Australian developments that suggest an emerging maturity in mathematics education. Australian educators appear to be having a greater impact than ever before, and confidence has stemmed from the greater opportunities now available to Australian mathematics educators to share ideas. Because there are now more professionally trained tertiary mathematics educators in Australia (Jones, 1984), it is not surprising that, at last, fresh thinking is seeking to identify the peculiar needs of Australian mathematics students and teachers.

References

Ashby, E. (1942). Our tenpenny universities. August 1, 1942. *The ABC Weekly*, pp. 3, 4.

Auchmuty, J. (1980). *Report of the National Inquiry into Teacher Education*. Canberra: Australian Government Publishing Service.

Australian Association of Mathematics Teachers Inc. [AAMT] (2002). *Standards for excellence in teaching mathematics in Australian schools*. Adelaide: Author.

Australian Bureau of Statistics (2017). http://www.abs.gov.au/ausstats/abs@.nsf/0/68AE74ED632E17A6CA2573D200110075?opendocument

Australian Bureau of Statistics (2001). *1301.0 - Year Book Australia, 200 Australian schools: Participation and funding 1901 to 2000*. http://www.abs.gov.au/Ausstats/abs@.nsf/0/A75909A2108CECAACA2569DE002539FB?Open

Australian Curriculum, Assessment and Reporting Authority [ACARA] (2009). Australian Curriculum—Mathematics (reprinted several times, with version 8.2 being the current version. Available from http://www.australiancurriculum.edu.au/download/f10

Australian Curriculum, Assessment and Reporting Authority [ACARA] (2016). Australian Curriculum: cross curriculum priorities. http://v7-5.australiancurriculum.edu.au/CrossCurriculumPriorities/Aboriginal-and-Torres-Strait-Islander-histories-and-cultures

Australian Curriculum, Assessment and Reporting Authority [ACARA] (2014). National assessment program—literacy and numeracy achievement in reading, persuasive writing, language conventions and numeracy: National report for 2014. Sydney: ACARA.

Australian Government Office of the Chief Scientist (2014). *Science, technology, engineering and mathematics: Australia's future*. Canberra: Australian Government.

Australian Institute for Teaching and School Leadership (2011). Accreditation of initial teacher education programs in Australia: Standards and procedures. Canberra: MCEECDYA.

Barnes, M. (1991). *Investigating change: An introduction to calculus for Australian schools* [Electronic resource]. Carlton South, Vic: Victorian Curriculum Corporation. http://trove.nla.gov.au/version/208131918

Bartlett, L., Knight, J., Lingard, B., & Porter, P. (1994). Redefining a national agenda in education: The states fight back. *Australian Educational Researcher, 21*(2), 29–44.

Becker, W. C. (1992). Direct instruction: A twenty year review. In R. P. West, & L. A. Hamerlynck (Eds.), *Designs for excellence in education: The legacy of B. F. Skinner* (pp. 71-112). Longmont, Colorado: Sopris West, Inc.

Bishop, A. J. (October 9-11, 1992). *Removing cultural barriers to numeracy.* Plenary address at the National Conference of the *Australian* Institute for *Teaching* and School Leadership, October 9–11, Sydney.

Biggs, J. B., & Collis, K. F. (1982). *Evaluating the quality of learning—the SOLO taxonomy.* New York: Academic Press.

Bishop, A. J. (1990). *Race & Class, 32*(2), 51–65.

Blackburn, J. (1985). *Ministerial review of postcompulsory schooling report: vol. 1* [Blackburn report]. Melbourne: The Review.

Blake, L. J. (1973). Vision and realisation: A century history of state education in Victoria. Melbourne: Education Department of Victoria.

Blakers, A. L. (1978). Change in mathematics education since the late 1950s: Ideas and realisation. *Educational Studies in Mathematics, 9*(2), 147–58.

Board, P. (no date). *Government schools of New South Wales from 1848: Reform movement.* http://www.governmentschools.det.nsw.edu.au/story/reform_movement

Brook, J. & Kohen, J. L. (1991). The Parramatta Native Institution and the Black Town: A history. Sydney: UNSW Press.

Burkhardt, G. (no year). *The first school teachers and schools in colonial New South Wales 1789–1810.* Australian National Museum of Education, anme.org.au/wp-content/uploads/2014/08/ANME_First_Teachers.pdf

Cadzow, A. (2008). *A NSW Aboriginal education timeline 1788–2007.* Retrieved from http://ab-ed.boardofstudies.nsw.edu.au/files/timeline1788-2007.pdf.

Cavanagh, M. (1996). Student understandings in differential calculus. In P. C. Clarkson (Ed.), Technology in mathematics education: Proceedings of the 19th annual conference of the Mathematics Education Research Group of Australasia (pp. 107–114). Melbourne: MERGA.

Christie, M. (1979). *Aborigines in colonial Victoria 1835–86.* Sydney: Sydney University Press.

Clarke, E. (1987). Assessment in Queensland secondary schools: Two decades of change 1964-1983. (Historical Perspectives on Contemporary Issues in Queensland Education No. 4). Brisbane: Policy and Information Services Branch, Department of Education, Queensland.

Clements, M. A. & Ellerton, N. F. (1996) *Mathematics education research: Past, present and future.* Bangkok, Thailand: UNESCO.

Clements, M. A., Grimison, L., & Ellerton, N. F. (1989). Colonialism and school mathematics in Australia 1788–1988. In N. F. Ellerton & M. A. Clements (Eds.), *School mathematics: The challenge to change* (pp. 50–78). Geelong, Vic: Deakin University.

Cockroft et al. (1982). Mathematics counts: Report of the Committee of Inquiry into the Teaching of Mathematics in Schools under the Chairmanship of Dr WH Cockcroft [The Cockcroft Report]. London: Her Majesty's Stationery Office.

Cohen G. L. (2006). Counting Australia in: the people, organisations and institutions of Australian mathematics. Broadway Bay, NSW: Halstead Press.

Commonwealth of Australia (1997). Bring them home: Report of the national inquiry into the separation of Aboriginal and Torres Strait Islander children from their families. Canberra: Human Rights and Equal Opportunity Commission.

Committee of Inquiry into education in Western Australia (K. Beasley, Chair). (1984). *Education in Western Australia: Report.* Perth: WA Government Press.

Committee for the Review of Tertiary Entrance Scores in The Australian Capital Territory (1986). *Making Admission to Higher Education Fairer: Report.* Canberra: Australian Capital Territory Schools Authority.

Commonwealth Schools Commission (1987). In the national interest: Secondary education and youth policy in Australia. Canberra: The Commission.

Crane, A. R. (1950). Student opinions of secondary courses, especially in Mathematics and Latin. *Forum of Education, IX*(I), 36–43.

De Bortoli, L., & Thomson, S. (2010). *Contextual factors that influence the achievement of Australia's Indigenous students: Results from PISA 2000–2006.* Retrieved from http://www.acer.edu.au/documents/pisa-indigenous-contextual-factors.pdf.

De Maio, J. A., Zubrick, S. R, Silburn, S. R., Lawrence, D. M., Mitrou, F. G., Dalby, R. B., Blair, E. M., Griffin J., Milroy, H., & Cox, A. (2005). *The Western Australian Aboriginal child health survey: Measuring the social and emotional wellbeing of Aboriginal children and intergenerational effects of forced separation.* Perth: Curtin University of Technology and Telethon Institute for Child Health Research.

Department of Education, Tasmania (2016). *2016 Course information handbook: For years 11 and 12.* Hobart: Tasmanian Department of Education, Tasmanian Government.

Donnelly, K., & Wiltshire, K. (2014). *Review of the Australian curriculum: Final report.* Canberra: Australian Government Department of Education.

Dyson, M. (2005). Australian teacher education: Although reviewed to the eyeball is there evidence of significant change, and where to now? *Australian Journal of Teacher Education, 30*(1), 37–54.

Ellerton, N. F. & Clements, M. A. (1988). Reshaping school mathematics in Australia 1788–1988, *Australian Journal of Education, 32*(3), 387–405.

Ewing, B. (2011). Direct instruction in mathematics: Issues for schools with high indigenous enrolments: A Literature Review. *Australian Journal of Teacher Education, 36*(5), 65–92.

FitzSimons, G. E., Seah, W. T., Bishop, A. J., & Clarkson, P. C. (2000). Conceptions of values and mathematics education held by Australian primary teachers: Preliminary findings from VAMP. In W.-S. Horng & F.-L. Lin (Eds.), *Proceedings of the HPM*

2000 conference, History in mathematics education: Challenges for the new millennium (Vol. 2), (pp. 163–171). Taipei: National Taiwan Normal University.

Fletcher, J. J. (1989). Clean, clad and courteous: A history of Aboriginal education in NSW. Carlton, NSW: J. J. Fletcher.

Forgasz, H. J. (2006). Australian year 12 mathematics enrolments: Patterns and trends—past and present. The University of Melbourne International Centre of Excellence for Education in Mathematics. http://amsi.org.au/publications/australian-year-12-mathematics-enrolments-patterns-trends-past-present/

Garner, S. (2005). Safely does it with CAS: Where have we been, where are we going? In M. Coupland, J. Anderson & T. Spencer (Eds.), *Making mathematics vital: Proceedings of the Twentieth Biennial Conference of The Australian Association of Mathematics Teachers* (pp. 120–132). Adelaide: AAMT.

Grant, S. (2016). *Stan Grant's speech on racism and the Australian dream.* (Foxtel IQ2 debate, recorded 2015, broadcast 24 January, 2016). Sydney: The Ethics Centre. http://www.ethics.org.au/on-ethics/blog/january-2016/stan-grant-s-speech-on-racism-and-the-australian-d

Geiger, V., Faragher, R., & Goos (2010). CAS-enabled technologies as 'agents provocateurs' in teaching and learning mathematical modelling in secondary school classrooms. *Mathematics Education Research Journal, 20*(2), 48–68.

Godfrey, C. (1912). The algebra syllabus in the secondary school. In International Commission on the Teaching of Mathematics: The teaching of mathematics in the United Kingdom. Part 1. London: HMSO.

Gravemeijer, K., & Doorman, M. (1999). Context problems in realistic mathematics education: a calculus course as an example. *Education Studies in Mathematics, 39*(1–3), 111–129.

Grimison, L. (1982). Historical aspects of the development of a mathematics curriculum in N.S.W. secondary schools to 1960. In J. Veness (Ed.), *Mathematics: A universal language* (pp. 39–45). Sydney: Australian Association of Mathematics Teachers.

Horwood, J. (1997). Professionalisation and change in secondary mathematics. In F. Biddulph & K. Carr (Eds.), People in mathematics education: Proceedings of the 20th annual conference of the Mathematics Education Research Group of Australasia (pp. 224–230). Rotorua, New Zealand: MERGA.

Hoy, A. (1957). Present trends and problems in secondary education. *Australian Journal of Education, 1*(1), 3–14.

Hughes, P. W. (1969). Changes in primary curriculum in Tasmania. *Australian Journal of Education, 13*(2), 130–146.

Hyams, B. K. (1979). *Teacher preparation in Australia: A history of its development from 1850 to 1950.* Hawthorn, Vic: Australian Council for Educational Research.

Ingvarson, L., Kleinhenz, E., & Wilkinson, J. (2007). *Research on performance pay for teachers.* http://research.acer.edu.au/workforce/1

Lean, G. A. (1994). *Counting systems of Papua New Guinea and Oceania* (Volumes 1–10). PhD thesis, University of Technology, Lae, Papua New Guinea.

Lamb, S. (1998). Completing school in Australia: Trends in the 1990s. *Australian Journal of Education, 42*(1), 5–31.

Lingard, B., Porter, P., Bartlett, L., & Knight, J. (1995). Federal/state mediations in the Australian national education agenda: From the AEC to MCEETYA 1987–1993. *Australian Journal of Education, 39*(1), 41–66.

Lyons, T., Cooksey, R., Panizzon, D., Parnell, A., & Pegg, J. (2006). *Science, ICT and mathematics education in rural and regional Australia: The SiMERR national survey.* Armidale, NSW: University of New England.

MacLeod, R (Ed.) (1988). The commonwealth of science: ANZAAS and the scientific enterprise in Australia 1888–1988. Melbourne: OUP.

Masters, G. N., & Hill, P. W. (1988). Reforming the assessment of student achievement in the senior secondary school. *Journal of Education, 32*(3), 274–286.

Matthews, C. (2015). RAMR cycle. http://ydc.qut.edu.au/documents/RAMR-with-Matthews-framework.pdf

Mays, H. (2005). Mathematical knowledge of some entrants to a pre-service education course. In M. Coupland, J. Anderson & T. Spencer (Eds.), *Proceedings of the 20th Biennial Conference of the Australian Association of Mathematics Teachers* (CD-ROM) (pp. 186–193). Sydney: AAMT.

Morony, W. (2010). AAMT response the draft national professional standards. Adelaide: AAMT.

Mousley, J., Sullivan, P., & Zevenbergen, R. (2004). Alternative learning trajectories. In I. Putt, R. Faragher, & M. McLean (Eds.), Mathematics education for the third millennium: Towards 2010*: Proceedings of the 27th annual conference of the Mathematics Education Research Group of Australasia, Townsville* (pp. 374–381). Sydney: MERGA.

National Assessment Program [NAP]. *Home page*. http://www.nap.edu.au/

National Council of Teachers of Mathematics (1980). *An agenda for action.* Reston, VA: NCTM.

NSW Department of Education (1980). *Sydney and the bush: A pictorial history of education in New South Wales* (compiled by J. Burnswoods and J. Fletcher for the NSW Department of Education). Sydney: N.S.W. Dept. of Education.

NSW Department of Education. Government schools of New South Wales from 1848. Sydney: NSW Education & Communities. http://www.governmentschools.det.nsw.edu.au/facts_figures.shtm

Ormell, C. (1972). Mathematics, applicable versus pure and applied, *International Journal of Mathematical Education in Science and Technology*, 3(2), 125–131.

Partridge, P. H. (1968). *Society, schools and progress in Australia,* Oxford, UK: Pergamon Press.

Pearson, N. (n.d.) *Cape York Partnership: Education* (website). http://capeyorkpartnership. org.au/cogs-of-change/educational

Relton, W. J. (1962). The failure of the dual system of control of education in N.S.W. *Australian Journal of Education, 7*(2), 133–142.

Santiago, P., Donaldson, G., Herman, J., & Shewbridge, C. (2011). *OECD reviews of evaluation and assessment in education—Australia.* Paris: OECD.

Sarra, C. (Ed.) (2011). Strong and smart: Towards a pedagogy for emancipation: education for first peoples.

Selleck, R. J. W. (1990). *Biography—Frank Tate, 1864–1939. An Australian dictionary of biography* (vol. 12). Melbourne: Melbourne University Press. http://adb.anu.edu.au/ biography/tate-frank-8748

Shears, L. W. (1957). Recent research on Australian secondary education. *Australian Journal of Education, 1*(1), 27–40.

Sullivan, P., Bourke, D., & Scott, A. *(1995).* Open-ended tasks as stimuli for learning mathematics. In S. Flavel et al. (Eds.), *GALTHA: Proceedings of the 18th annual conference of the Mathematics Education Research Group of Australasia* (pp. 484–492). Darwin: MERGA.

Thomson, S., McKelvie, P. & Murnane, H. (2006). Achievement of Australia's early secondary indigenous students: Findings from TIMSS 2003 (TIMSS Australia Monograph No 10). Camberwell, Vic: The Australian Council for Educational Research.

The University of Melbourne (2008). *Melbourne: The city past and present: Education, prior to 1872.* http://www.emelbourne.net.au/biogs/EM00507b.htm

Turney, C. (Ed). (1969). *Pioneers of Australian education.* Sydney: Sydney University Press.

Varavsky, C. (2012). Use of CAS in secondary school: a factor influencing the transition to university-level mathematics? *International Journal of Mathematical Education in Science and Technology, 43*(1), 33–42.

Veness, J. (1985). *The Mathematical Association of NSW: An account of the first seventy-five years.* https://www.mansw.nsw.edu.au/about-us/mansw-association-history.

von Glasersfeld, E. (1990). An exposition of constructivism: Why some like it radical. In R. B. Davis, C. A. Maher, and N. Noddings (Eds.), *Monographs of the Journal for Research in Mathematics Education* (pp. 19–29). Reston, VA: National Council of Teachers of Mathematics.

Webb, L. C. (1963). The teaching profession. *The Australian Journal of Education, 7*(3), 145–156.

White, A. L., Way, J., Perry, B., & Southwell, B. (2005/6). Mathematical attitudes, beliefs and achievement in primary pre-service mathematics teacher education. *Mathematics Teacher Education and Development, 7,* 33–52.

White, M. A. (1975). Sixty years of public examinations and matriculation policy in Western Australia. *Australian Journal of Education, 19*(1), 64–77.

Wienk, M. (2015). Discipline profile of the mathematical sciences 2015. Melbourne: AMSI.

About the Authors

Dr Judith Mousley taught in pre-school, primary and secondary schools for fifteen years before joining Deakin University. She coordinated mathematics education courses in the Faculty of Education's undergraduate and postgraduate programs, and had a strong record of attracting research and development funds. Her publications include edited books, chapters, research reports, journal articles, videotapes and a CD. Judy was President of the Australian Mathematical Sciences Council, Vice President (Teaching) of the Mathematics Education Research Group of Australasia, Vice President of the International Group for the Psychology of Mathematics Education, and an executive member of a range of other professional organisations. Her higher degree students worked in the areas of philosophies of mathematics education, the history of education, curriculum change, use of technologies in education, and higher education policy.

Dr Chris Matthews is from the Quandamooka people of Minjerribah (Stradbroke Island) in Queensland Australia. Chris has received a PhD in applied mathematics from Griffith University and is a Senior Lecturer at the Griffith School of Environment, Griffith University. Chris has undertaken numerous research projects within applied mathematics and mathematics education. Chris was the patron and expert advisor for the Make It Count Project; a large mathematics education project coordinating education research within clusters of schools across Australia with the specific aim of improving mathematics education for Indigenous students. Currently, Chris is the chair of the Aboriginal and Torres Strait Islander Mathematics Alliance (ATSIMA) which aims to improve educational outcomes in mathematics for Aboriginal and Torres Strait Islander learners. Chris was also awarded the STEM Professional Award under the 2016 Indigenous STEM Awards awarded by BHP Billiton Foundation and CSIRO for his work in STEM education.

Chapter V

PHILIPPINES: Mathematics and its Teaching in the Philippines

Ester B. Ogena
Philippine Normal University, Manila

Marilyn Ubiña-Balagtas
Philippine Normal University, Manila

Rosemarievic V. Diaz
Philippine Normal University, Manila

Introduction

Mathematics is a core subject across levels and programs in the Philippine education system. It is a discipline whose language is unique and precise. It is a tool to one's daily life. Its importance to individuals and the country's development is unquestionable. Its basic skills are needed across time and space. Its truthfulness and applications last for generations. However, some learners view Mathematics as only meant for high achievers. Mathematics has then developed a good and bad reputation among learners. Its value is undoubtedly recognized but how it is taught has become a recurring issue.

In the Philippines, the way mathematics is taught in basic education is a concern of the Department of Education (DepEd). To DepEd (2013), mathematics is one subject that pervades life at any age and in any circumstance. Its value goes beyond the classroom and the school. As a school subject, it should therefore be learned comprehensively and with much depth.

On the other hand, the way mathematics is taught is a concern of the Commission on Higher Education (CHED). This is under mathematics education, which is a program that develops and strengthens pre-service and in-service teachers' content knowledge and pedagogy in teaching the discipline. In the Philippines, CHED sets the national curriculum standards in mathematics and mathematics education that all Higher Education Institutions (HEIs) should meet for uniformity in desired learning outcomes and for harmony in teaching practice. Variability may happen in the way the program is delivered. Nevertheless, for quality assurance, the program should develop Filipino learners with high achievement and productivity, a quality life, and a pride in their country.

In the history of the Philippines, the K to 12 Program is a major education reform. This reform is not only a concern of DepEd and CHED but of all agencies that receive the graduates of HEIs. The transformation in basic education from the usual 10 to 13 years is worth examining. Its feature of compulsory 2-year senior high school gives hopes of making Filipino citizens competitive and respected by professionals across the globe. Given the major change in the design of mathematics in the basic education curriculum from discipline-based to spiral progression, it is good to study how this evolved across time. It is at this juncture that this chapter has been conceived. Its purpose is to trace the beginnings and developments that took place in the teaching of mathematics in schools and HEIs in the history of Philippine education.

However, the writers deemed it necessary to acknowledge the scarcity of their primary sources as there are limited resources that can be referred to about the history of mathematics education in the Philippines. Nevertheless, they have supported their work with their personal experiences as learners and their professional journey as teacher educators, researchers and leaders in promoting mathematics and mathematics education in the country.

History of Mathematics and Mathematics Education in the Philippines

Pre-colonial Period

Filipino historians have difficulty describing the history of the Philippines, particularly in the pre-colonial period. Hence, not much can be written specifically on how mathematics and mathematics education were used and taught in the education of the Filipino people. Nevertheless, reports of anthropologists about remnants of people who lived in the Philippine archipelago and writings found in caves and trees could reveal the kind of mathematics learned and applied during the pre-colonial period. This section then attempts to describe the mathematics that early Filipinos learned and applied in prehistoric period ending in 900 AD and pre-Spanish period ending in 1521 (Wikipedia, 2016).

Pre-historic Period

The pre-historic period covers events occurring before April 21, 900, the date of the Laguna Inscription, currently the oldest known written record relating to the Philippines. This period is traced back to the *Stone Age* culture around 50,000 to 500 BC. The first Filipino anthropologist, Felipe Landa Jocano, explained that stone tools and ceramic manufacture are the two core industries that defined the period's economic industry (Wikipedia, 2016, para 1). Inferred from this account, the mathematics applied in this period focused on skills in the production of stone tools and ceramics for trade and industry (e.g. calculation, estimation, measurement and geometry).

Then the *"Tabon Man" period* came with the discovery of fossilized fragments of skull and jawbones at the Tabon Cave in Palawan, estimated to have lived in about 24,000 to 22,000 BC. The American anthropologist, Dr. Robert B. Fox, discovered these fossils on May 28, 1962 and believed that the cave was a kind of *Stone Age* factory. Part of the Pre-historic time is the *Iron Age*, believed to have existed during the 9th and 10th centuries BC. Then came the *Metal Age* in about 500 BC to 1 AD. The artefacts found in different sites throughout the Philippine archipelago show similar designs indicative of the use of metal tools and pottery technology of a group of people in this period (Wikipedia, 2016). Given the nature of the metal and pottery industry, early Filipinos must have been good in calculating and estimating, part of the language of trade and pottery industry.

According to Manapat (2011), there were basic geometric ideas found in the Angono Petroglyphs, a set of pre-historic rock and cave drawings found in the hills of Angono, a mountainous area to the South of Manila, conjectured to have existed in the late Neolithic Period, about 3,000 BC. The geometric figures formed demonstrate notions of symmetry, proportion, and mathematical scaling as seen in the successful resizing of stone etchings.

There were also other theories that have emerged to explain the life of the pre-colonial people in the Philippines. One is the *Wave Migration Theory* of H. Otley Beyer, explaining that the first Filipino ancestors came to the island in different waves of migration. Some came through land bridges when the sea level was low or through water vessels called *balangay*. These migrations explain that Filipino ancestors were good at calculating distance, mass and time.

Pre-Spanish Period

The pre-Spanish Period started upon the discovery of the Laguna Copperplate Inscription (LCI) in 900. The LCI is the first written document found in the Philippine language, which was dated to have been written in 900 but discovered only in 1989 (Wikipedia, 2016). The LCI states that "the descendants of *Namwaran* are forgiven from a debt of 926.4 grams of gold, and is granted by the chief of Tondo (an area in Manila) and the authorities of Paila, Binwangan and Pulilan (places in Luzon). The words are a mixture of mostly Sanskrit, Old Malay, Old Javanese and Old Tagalog. The writings reveal that early societies existed in the Philippines with traders and rulers doing international trading. This trading signifies use of standard units of mass measurement up to grams and decimals by early Filipinos, before the colonial period.

Colonial Period

This section covers the colonial period during the Spanish (1521–1898), American (1898–1946) and Japanese (1941–1945) colonization of the Philippines.

Spanish Colonial Era

The Spaniards colonized the Philippines for more than three centuries. The colony began when the European explorer, Ferdinand Magellan, landed in the Philippines on March 16, 1521 and claimed the island under the possession of King Philip of Spain. During Spain's rule of the Philippines, the economy depended mostly on the Galleon Trade. Manila became the center of trade. Taxation was introduced. Government buildings, churches, schools and markets were built around plazas. The indigenous population in Manila was reduced or relocated. Native traditions and practices were transformed into Spanish culture. Education was mostly religion-oriented with the friars and missionaries teaching Catholicism, reading, writing, industrial and agricultural techniques using the natives' local language. Primary instruction was made free and open to all natives regardless of race and resources. The teaching of Spanish was made compulsory.

A royal decree provided a complete educational system to the Filipinos, consisting of primary, secondary and tertiary levels. Special schools for nautical, medical, pharmaceutical, accounting, vocational, agriculture and French and English Languages were established. In secondary schools, a range of courses were taught to include mathematics such as arithmetic, algebra, geometry, and trigonometry. Tertiary education began in 1611 with the establishment of college seminaries whose mission was for social refinement and distinction (Ogena, Brawner, & Ibe, 2013). Tertiary education was extremely prestigious but only the sons of Spaniards and elite Filipinos could afford it. Courses in this level include advanced mathematics, agriculture, chemistry, commerce, geography, topography, history, mechanics, painting, philosophy, physics, rhetoric, poetry, and languages such as Spanish, English, French, Greek, and Latin.

The opening of the Suez Canal in 1869 made traveling to Spain easier and more affordable to many Filipinos who wanted to pursue higher education. This increased the number of educated Filipinos in Spain and Europe, which later led to the formation of an enlightened social class called *Ilustrados*. This group led the Philippine independence movement and used the Spanish language as its main communication method. The most prominent Filipino in this group was Jose Rizal, now the Philippine's national hero. He inspired Filipinos to aim for independence through his novels written in Spanish, which led to his conviction and public execution on December 30, 1896 at the age of 35. Then several revolutionary movements took place that led to the proclamation of the first Philippine Republic on June 12, 1898.

American Colonial Era

The Americans colonized the Philippines for more than four decades. The signing of the Treaty of Paris between the United States (US) and Spain in December 1898 marked the end of Spanish colonization of the Philippines and start of the American regime. During the American period, there were also three levels of education: elementary, secondary and tertiary. Elementary education was free and started at age 7, with four years at the primary and three years at the intermediate level. Secondary school was good for four years. Religion was not part of the curriculum. New schools, such as normal, vocational, agricultural and business schools were opened for college education. Scholarships were also introduced for those who excelled academically. They were sent to the United States to study and later became leaders in the government.

In 1901, Act 74, also known as Education Act 1901, established the Department of Public Instruction in the Philippine Islands and a normal and trade school in Manila. The first Filipino teachers in these schools were volunteer American soldiers who were later replaced by more qualified American teachers called Thomasites, a name derived from the ship they traveled on. They were deployed in Manila and in some provinces in the Philippines. The Filipinos were taught first in their language before English was made a common medium of communication due to many variants of Filipino dialects. This period also gave rise to a profound public/private split in the Philippine education system as more private sectarian and non-sectarian colleges were established (Ogena, Brawner, & Ibe, 2013).

The first normal school for teachers, established on September 1, 1901, was the Philippine Normal School (PNS), now the Philippine Normal University (PNU), the Alma Mater and work station of the writers. Its establishment marked the time when teacher education began to take shape (Ogena, Brawner, & Ibe, 2013).

According to the University of the Philippines, National Institute for Science and Mathematics Education Development (UP-NISMED, 2001), education during the American period aimed at preparing individuals to exercise the right for suffrage intelligently and to perform the duties of citizenship fully and honestly. The 3Rs (Arithmetic, Reading and Writing) were emphasized in the primary level. The intermediate curriculum aimed to provide training for useful occupations like farming, trade, housekeeping, household arts, and business. Secondary education prepared students to pursue university education related to clerical, trade or agriculture work. In terms of the teaching of Mathematics, most textbooks were co-authored by Filipino educators with Americans. Sand tables were used as an instructional device in all subject areas including Mathematics. In terms of teaching strategy, memorization of mathematical facts and problem solving were emphasized. The teaching of Mathematics and Science was found to be the same as their counterparts in the USA, according to a report by the Monroe Commission on Philippine Education, created in 1925 to evaluate the American Curriculum in the Philippines.

Japanese Colonial Era

The Japanese occupied the Philippines during World War II, soon after Japan attacked Pearl Harbor on December 7, 1941. Japanese troops landed in the Philippines and occupied many places in Luzon, including Manila. The Japanese occupiers tried to establish an educational system that was Asian and nationalist in perspective, with emphasis on vocational, technical and science education in all levels to replace what the Americans had introduced (Ogena, Brawner, & Ibe, 2013 p. 162). Filipino soldiers, recognized as American nationals, fought against the Japanese in defense of the Americans and successfully ended the Japanese empire in the Philippines on September 2, 1945.

According to UP-NISMED (2001), during the Japanese Occupation, Mathematics was just considered a tool to achieve the objectives of a production-oriented curriculum. There was a lot of practical work and development of manipulative skills like measuring and drawing. The secondary curriculum then prepared Filipino citizens to become skilled workers and productive citizens. Japanese language was also introduced as a required subject.

Post-colonial Period

The post-colonial period started when the Philippines gained its independence from America on July 4, 1946, with the signing of the Treaty of Manila between the USA and Philippine governments. During this period, many private tertiary education institutions have been founded, as government funds were provided only for primary and secondary education. Tagalog, the national language, which later on became Filipino, was declared as one of the official languages. Mathematics was taught in English and became a part of the curriculum from elementary to higher education.

In 1957, the launching by the Soviet Union of the first satellite ('Sputnik') to orbit the earth changed the landscape of science education in the Philippines. Under the Revised Philippine Educational Program, the value of Science as well as Arithmetic was emphasized and these were taught daily for 40 minutes in lower primary and 50 minutes in higher primary. In high school, Mathematics was taught in two curricula: vocationally-inclined and college-bound. In the first two years, Mathematics was taught for 40 minutes in both curricula. In their third and fourth years, 80-minute and 40-minute Mathematics were taught in the college-bound and vocationally-inclined curriculum, respectively.

In the mid-60s, the new Mathematics Curriculum emphasized the logical way of thinking and the precision of Mathematics language. Mechanical computations were considered tools for problem solving. In this period, the Ministry of Science and Technology (now the Department of Science and Technology, DOST) established special Science high schools (e.g. Manila Science High School and Philippine Science High School in 1963) for highly select elementary graduates. In these schools, Mathematics, Science and English were given heavier weights compared with other subject areas. Mathematics courses included General Mathematics (Math 1), Plane and Solid Geometry (Math 2), Algebra (Math 3), Advanced Algebra and Trigonometry (Math 4), Elementary Analysis (Math 5), and Pre-Calculus in higher years.

In 1972, the declaration of Martial Law required a different orientation to the curriculum. Secondary school curriculum saw the strengthening of science, mathematics and vocational education as contributory to national development. Under this curriculum, arithmetic, algebra and geometry were prescribed from first to fourth year but the topics differed per year: Number System (1st year); Algebra (2nd year); Geometry (3rd year); and Advanced Algebra (4th year). It also prescribed electives in the 2nd year, such as Consumer Mathematics, Statistics and Trigonometry. In teacher education, pre-service teachers needed to take one science course in the 2-year Elementary Teacher Certificate (ETC) and two science courses for the Bachelor of Science in Elementary Education (BSEED). Then later, the ETC was replaced with the BSEED.

In 1983, the New Elementary School Curriculum (NESC) was implemented, prescribing mastery learning. Under this curriculum, mathematics was taught in the primary level for 40 minutes daily. In the lower primary, mathematics focused on the four fundamental operations, fractions, the metric system, local measurements, and applications of money in real-life problems. In the upper primary, mathematics emphasized ratio and proportion, angles, plane and spatial figures, scales, maps, graphs, practical business and industry related problems prevalent in the community (UP-NISMED, 2001). In 1989, the Department of Culture and Sports (DECS) implemented the Secondary Education Development Plan (SEDP). Under this curriculum, mathematics has been allotted from 180 to 200 minutes per week. Several reforms happened in the basic education curriculum, but basically, it was implemented only for 10 years.

In teacher education, mathematics teachers were trained through different degree programs offered in Teacher Education Institutions (TEIs), including Bachelor of Elementary Education (BEED) and Bachelor of Secondary Education (BSEd) with specialization in Mathematics. The PNU offered alternative programs like Bachelor of Science in Elementary and Secondary Education (BSESE) major in Mathematics and the Bachelor of Science in Mathematics for Teachers (BSMT).

From 1978 to 1995, after completing their program, trained teachers also had to satisfy the requirements of the Professional Board for Teachers (PBET), used by the Civil Service Commission (CSC) to certify teachers. In 1996, the Professional Regulatory Commission (PRC) replaced PBET with the Licensure Examination for Teachers (LET), a test that up to now is the main determinant for a teacher education degree graduate to be a qualified teacher.

Current Programs in Mathematics and Mathematics Education in the Philippines

Mathematics in basic education

Triggered by the poor performance of Filipino students in mathematics and science in international examinations and the fact that almost all countries in the world have 12-year basic education, DepEd designed its K to 12 Program. Under this program, Mathematics Curriculum has the twin goals of developing critical thinking and problem-solving skills among basic education learners. These goals are hoped to be achieved with an organized and rigorous curriculum content, a well-defined set of high-level skills and processes, desirable values and attitudes, and appropriate tools, considering the different contexts of Filipino learners (DepEd, 2013). Mathematics content from Grade 1 to 10 has five strands designed in spiral progression, namely: Numbers and Number Sense; Measurement; Geometry; Patterns and Algebra; and Probability and Statistics. *Numbers and Number Sense* includes concepts of numbers, properties, operations, estimation, and their applications. *Measurement* includes the use of measures like length, mass, weight, capacity, time, money, temperature, perimeter, area, surface area, volume, and angle measures to describe, understand, and compare mathematical and concrete objects. *Geometry* includes properties of 2- to 3-dimensional figures and their relationships, spatial visualization, reasoning, and geometric modelling and proofs. *Patterns and Algebra* studies patterns, relationships, and changes among shapes and quantities and the use of algebraic notations and symbols, equations, and most importantly, functions, to represent and analyze relationships. *Statistics and Probability* is all about developing skills in collecting and organizing data using charts, tables, and graphs; understanding, analyzing and interpreting data; dealing with uncertainty; and making predictions about outcomes. In Senior High School, the mathematics courses include General Mathematics and Statistics & Probability offered as core subjects. Then Pre-Calculus and Basic Calculus are offered in the academic track of Science, Technology, Engineering and Mathematics.

Moreover, DepEd (2013) emphasized that:

The specific skills and processes to be developed in Mathematics are knowing and understanding; estimating, computing and solving; visualizing and modelling; representing and communicating; conjecturing, reasoning, proving and decision-making; and applying and connecting. The values and attitudes to be honed should include accuracy, creativity, objectivity, perseverance, and productivity. Use of appropriate tools is necessary in teaching mathematics and these should include manipulative objects, measuring devices, calculators and computers, smart phones and tablet PCs, and the internet. The context of Mathematics is defined as a locale, situation, or set of conditions of Filipino learners that may influence their study and use of mathematics to develop critical thinking and problem solving skills. Contexts refer to beliefs, environment, language and culture that include traditions and practices, as well as the learners' prior knowledge and experiences.

Given the changes in the structure of mathematics and the enrichment in its content and delivery, including its openness to different technologies, it is hoped that the K to 12 Reform makes the quality of education in the Philippines globally competitive.

Mathematics in teacher education

In 1990, the government created the Congressional Commission on Education (EDCOM) to study, review and assess the Philippine Education. The EDCOM report led to the establishment of the Commission on Higher Education (CHED) in 1994 through Republic Act (RA) 7722 as a responsible body in regulating the operations of HEIs in the country. CHED issues policies, guidelines, and minimum standards for programs including teacher education. Its Technical Panel in Teacher Education (TPTE) prescribes the curricular content of the program. Through RA 7784, a Teacher Education Institution (TEI) may be designated as Center of Excellence (COE). As COE, a TEI can design its own teacher education curriculum that the TPTE approves.

At present, two pathways can prepare teachers in the country; a four-year degree, either BSEd or BEED, and a Bachelor's degree either in Science or in Arts with Certificate in Teaching Program (CTP). The BSEd and BEED programs have the three components: General Education (GE), Professional Education (PE), and the Specialization or Content courses. The CTP requires at least 18 units of PE for the non-education degree holders to qualify to take LET.

At the time of writing this chapter, the teacher education curriculum in the Philippines is still based on the CHED Memorandum Order (CMO) no. 30 s.2004 that defines the policies and standards for undergraduate teacher education program. In this memorandum, both BEED and BSEd majors in Mathematics have this distribution of units: 63 units of GE; 54 units of PE for BEED and 51 for BSEd, and 57 units for Content Courses for BEED and 60 units of Specialization Courses for BSEd. Each curriculum has 174 units where every unit had 18-hour academic course work. This curriculum still covers a 10-year basic education but addresses EDCOM report findings stating that the teacher education curricula were heavy in GE, specialization courses were not sufficient in content, and that field experience and practice teaching were not enough.

One feature of this 2004 curriculum was the increase of Mathematics courses. Prior to 2004, the GE of BEED and BSEd curricula had only nine units of mathematics. Since the kind of GE mathematics courses to offer were not prescribed, TEIs offered those that LET covered, namely: Basic Mathematics, College Algebra and Plane Euclidean Geometry. In 2004 curriculum, the BEED has 6-unit GE mathematics and 12 more units of mathematics to strengthen the content knowledge and pedagogy of the graduates when they teach the subject in higher primary. For the BSEd major in mathematics, the 36–48 units of mathematics courses in the old curriculum had increased to at least 60 units in the 2004 curriculum. Mathematics courses include intermediate and advanced algebra, plane geometry, trigonometry, spherical trigonometry, analytic geometry, probability, differential and integral calculus, business mathematics and statistics.

Another feature of the 2004 curriculum is the strengthening of the classroom management and facilitating skills of the pre-service teachers in the PE courses of BEED and BSEd. The PE component includes 12 units on concept/theory, 27 units on methods/strategies, six 1-unit field study courses, and 6-unit practice teaching. In each field study course, students are provided practical learning experiences to verify and reflect on the theories learned through actual classroom observations. In this curriculum, prospective teachers are expected to be more prepared when they do their practice teaching in schools for at least nine weeks.

Cognizant of the mandate of COEs to innovate and design alternative models of the teacher education curriculum upon CHED's approval, PNU, a COE and the designated National Center for Teacher Education (NCTE) through RA 9647, designed and implemented an Outcome-Based Teacher Education Curriculum (OBTEC), implemented starting in school year 2014–2015. This curriculum is in synch with CHED's mandate for Outcomes-based Education in HEIs (CHED, 2012) and K to 12 compliant as well. In this OBTEC, PNU programs for pre-service teachers in Mathematics include: 1) Bachelor in Elementary Mathematics and Science Education (BEMSE), intended for those who will teach mathematics and science in higher primary levels; and 2) Bachelor in Mathematics Education (BME) with certificate program in teaching Junior

High School and Senior High School to produce teachers in the secondary level. The teaching of mathematics in kindergarten to Grade 3 is covered in the Bachelor of Early Childhood Education. The OBTEC has been designed based on the results of the International Association for the Evaluation of Educational Achievement (IEA) Teacher Education Development Study in Mathematics (TEDS-M), which sampled BEED and BSEd students in mathematics in the Philippines (Tatto et al., 2012).

Mathematics Performance of Filipino Learners

Mathematics performance in national assessments

The DepEd has been giving the National Achievement Test (NAT) in basic education since 2006. NAT replaced the National College Entrance Examination (NCEE) introduced in 1973 that graduating high school students had to take to know if they could go to college or vocational courses as well as the national assessments introduced in 1994 (i.e. National Elementary Achievement Test taken by graduating Grade Six pupils and the National Secondary Achievement Test taken by the fourth-year high school students). The NAT which is given to four levels (Grades 3, 6, Year 2 and 4) is taken by all public and randomly selected private schools.

According to Benito (2010), in the *NAT for Grade 3* (covering Science, Mathematics, English and Filipino), results in five school years (SY) beginning SY 2007–2008 showed that pupils tested had better foundation in Mathematics compared with other subject areas. In *NAT for Grade 6* covering the same four courses including Geography, History and Culture, the results showed that the mathematics performance of Grade Six pupils tested also in five school years had an increasing trend in all years, except in 2008–2009, indicative of improvement from *average mastery* to *moving towards mastery* in mathematics. In *NAT in Second Year High School* covering the same four subject areas including Social Studies, the performance in three school years beginning 2007–2008 indicates *average mastery* in mathematics consistently across years. In *NAT in Fourth Year*, students' performance in mathematics was consistently higher than the over-all test results of NAT in the first two school years of its implementation. An examination of the performance in mathematics of the students tested across levels showed no significant improvement across years and failure to meet the targeted criterion of mastery of at least 75% mean percentage score.

In addition to NAT, there were also national competitions in mathematics that the Mathematics Teachers Association of the Philippines (MTAP) organized with DepEd that were participated in by selected students in elementary and high schools. The MTAP competitions usually became the training ground for those who joined international competitions in mathematics.

Mathematics performance in international assessment

The Philippines joined the Trends in International Science and Mathematics Survey (TIMSS) in 1995, 1999 and 2003 (Ogena, Brawner, & Ibe, 2013, p. 168). In 2003, 95% of Filipino grade four pupils, excluding those with intellectual and functional disabilities, participated in TIMSS. The students have earned a standard mathematics average score of 358, which was below the international average score of 495 of 25 participating countries (Gonzales et al., 2004). The Philippines ranked third from the bottom outperforming only two countries (i.e. Tunisia and Morocco). The test covered *Numbers* (Patterns, Equations, Relationships); *Measurement*; *Geometry*; and *Data*. In the same year, 96% of Filipino Grade 8 population participated also in TIMSS and earned an average scale score of 378 in mathematics as against the international average of 466. This score shows a significant improvement when compared with the score of 345 obtained in 1999. The TIMSS in Grade 8 mathematics covered the same areas as that of Grade 4.

In 2008, 4,091 students from 118 science high schools and some elite private institutions in the Philippines participated in the TIMSS-Advanced that IEA organized. According to Ogena, Laña and Sasota (2010), TIMSS-Advanced assessed the performance of students in their final year of secondary schooling with special preparation in advanced mathematics and physics. The test was designed to cover two dimensions: content areas and cognitive domains. The content areas covered include algebra, calculus, and geometry. The cognitive domains covered include knowing, applying, and reasoning. *Knowing* pertains to the students' understanding of mathematical facts, concepts, tools, and procedures. *Applying* refers to the students' ability to make use of knowledge and conceptual understanding through problem solving situations. *Reasoning* goes beyond the solution of routine problems to cover unfamiliar situations, complex contexts, and multi-step problems. In this examination, the Philippine students placed at the bottom out of 10 participating countries with an average score of 355 as against the average scale score of 500.

The data revealed that among the three content areas, the students performed better in geometry than in calculus, which was their weakest. Among the three domains, they were better in knowledge compared with application and reasoning domains. In general, only 13% of the Filipino students reached the *intermediate benchmark*, a level with a score of at least 475, indicating that the students demonstrate knowledge of concepts and procedures in algebra, calculus and geometry to solve routine problems. Only 4% reached a *high benchmark* with an average score of at least 550, indicating that students can use their knowledge of mathematical concepts and procedures in algebra, calculus, and geometry and trigonometry to analyze and solve multi-step problems set in routine and non-routine context. Only 1% of the sampled students reached the *advanced benchmark* with an average score of at least 625, indicating that students can demonstrate their understanding of concepts, mastery of procedures, and mathematical reasoning skills in algebra, trigonometry, geometry, and differential and integral calculus to solve problems in complex contexts. Among the participating science and special schools, the DOST-supervised PSHS performed the best and can compete internationally in Mathematics (Ogena, Laña, & Sasota, 2010).

Other international examinations in mathematics for basic education learners

The Philippines also joined international competitions in mathematics. A very active group that trains students to join international examinations in mathematics is the Mathematics Trainers' Guild, Philippines (MTG). The MTG, created in 1995, is a non-profit organization of mathematics teachers committed to develop and promote excellence in mathematics education and training in the Philippines. Originally, MTG members were private schools in Metro Manila only, but through annual seminars, workshops and training, membership grew in number to include students from public schools. MTG institutionalizes excellence in mathematics education and training in the Philippines to upgrade the discipline to meet international standards and to provide the basis for scientific and technological manpower needed for the country's education, modernization and industrialization. Over the years, from 1996 to present, it has won many competitions in Asian cities in mathematics for primary and high school learners. For example, in 1996, it has brought home six bronze medals each for individual calculation skills and problem solving skills in primary mathematics competitions that were held in Hong Kong. Then in subsequent years, the students in the primary and secondary level consistently brought home several bronze, silver and gold medals from joining international competitions in Mathematics held in Hong Kong, China, Taiwan, USA, Canada, Australia, Korea, South Africa, India, Thailand, Indonesia, Malaysia, and Singapore, and Romania (MTG, 2016).

Mathematics performance of pre-service teachers in international examinations

The Philippines has also participated in the TEDS-M conducted in 2008. TEDS-M is the first international study involving future primary and lower-secondary school mathematics teachers, aimed at providing input to the current mathematics teacher education curriculum (Tatto et al., 2012). The study identified how the participating countries prepare their mathematics teachers and the variations in the nature and impact of teacher education programs. Although not the main intent of the study, the study revealed the need to strengthen the content knowledge and pedagogical content knowledge in mathematics of future primary and lower secondary mathematics teachers tested in the Philippines, having obtained a mean score below the benchmark of the 17 participating countries. Such results indicate the pre-service teachers' lack of readiness to deliver mathematics at the primary and lower-secondary level (Tatto, et al., 2012, pp. 129–149).

Going through the intentions of the national and international examinations that the Filipino students and pre-service teachers in mathematics have participated in, results reveal the need to strengthen mathematics education in the country. Although international competitions have shown Filipino students bringing home bags of medals and awards, the challenge remains for the Philippines to show its competitiveness given its major education reform at present.

Current Initiatives in Mathematics and Mathematics Education in The Philippines

Enhancement programs in mathematics

There are efforts to develop students' mastery in mathematics to have a strong foundation when taking up mathematics-related degree programs in college. According to DepEd Advisory No. 175 s. 2016, MTAP has programs for regular and talented students in mathematics to prepare the Grades 6 and 10 for entrance examinations for Grade 7 Junior High School (JHS) and Grade 11 Senior High School (SHS), respectively. The first program for the regular students aims to:

1. provide the learners with the opportunity to explore mathematics without the threat of tests; and
2. review the materials covered in the previous school year to enable them to do well in mathematics in the present school year.

The second program for the talented students in mathematics also aims to prepare the students for the Metrobank Foundation, Inc. (MBFI), MTAP and DepEd Math Challenge (MMC). Since 1997, the MMC takes place yearly and is aimed at contributing to the improvement of the quality of mathematics education in the Philippines (MMC, 2016). It is a math contest for elementary and secondary students in both public and private schools nationwide aimed to: a) awaken greater interest in mathematics; b) encourage students to strive for excellence in mathematics; c) encourage mastery of basic mathematical skills; d) discover mathematical talents among elementary and high school students; and e) provide students opportunities in leadership and cooperative undertaking.

Directions of national assessments covering mathematics under the K to 12 Reform

Under the K to 12 Reform, DepEd has set to administer several national assessments for different purposes (DepEd Order. No. 55, s. 2016). In all these national assessments, numeracy or mathematics is covered. As explained in this policy, these national examinations include: 1) Early Language, Literacy and Numeracy Assessment that will be administered at the end of Grade 3 as a key stage assessment to determine if students are meeting the learning standards in language, literacy and numeracy; 2) Exit Examinations to be administered in Grade 6, 10 and 12 to determine if students are meeting the learning standards in elementary, junior high school and senior high school curriculum; and 3) Career Assessment to be administered in Grade 9 to determine learners' aptitudes and occupational interests for career guidance. DepEd has further explained that these national assessments will remain in force and in effect until SY 2023–2024, which is the year when the first K to 12 cohort completes Grade 12.

System assessment under the K to 12 Reform

In assessing the effectiveness of the current education system and its impact on early grades not only on reading but also on their mathematical ability, a national assessment in mathematics will be administered where selected mother tongues are to be used. As described in the DepEd Order No. 57. s. 2015, the Early Grade Math Assessment (EGMA) is an individually administered oral test aimed at measuring the primary numeracy and mathematics skills of children in the early grades. DepEd adapted this examination following international standards and guidelines. The test is administered in selected mother tongues as a component of system assessment covering oral and rational counting, number identification and discrimination, missing number, addition, subtraction, word problems, and geometric pattern completion and visualization (DepEd, 2016, p. 3–4).

Moreover, DepEd has also planned participation in international examinations in the future (DepEd, 2015). Part of the plan of the Technical Working Group on System Assessment that DepEd formed in 2015 is to investigate future participation by the Philippines in selected international assessments. In the pipeline is its participation in the following international examinations where the importance of mathematics as part of the test is again challenged:

1. Southeast Asia-Primary Learning Metrics (SEA-PLM), which the Southeast Asian Ministers of Education Organisation (SEAMEO) leads to assess reading, writing, mathematics and global citizenship targeted for Grade 5 pupils of participating countries;

2. TIMSS organized by IEA that will cover major content and cognitive domains in four international benchmarks; and

3. Programme for International Student Assessment (PISA), led by the Organisation for Economic Cooperation and Development (OECD), given every three years, which tests the knowledge and skills in the areas of reading, science and mathematics of 15-year old students who are approaching the end of their compulsory education (DepEd, 2015).

Trends in Mathematics and Mathematics Education in The Philippines

In the period of the K to 12 Reform, several trends in the teaching, learning and assessment of mathematics have become apparent. Given the spiral and contextualized design of the mathematics curriculum, which replaces the discipline-based approach to teaching, particularly at the secondary education level, several approaches have emerged.

Teaching approaches in mathematics

In terms of teaching approach, a spiral progression approach in the design of the mathematics curriculum is used to ensure mastery in knowledge and skills in teaching mathematics. In this curriculum design, teaching approaches such as experiential, situated, reflective, constructivist, cooperative, discovery and inquiry-based teaching approaches must be strengthened. According to DepEd (2013, p. 4), *Experiential Learning* occurs by making sense of direct everyday experiences. *Situated Learning* is learning in the same context in which concepts and theories are applied. *Reflective Learning* refers to a deeper learning, which occurs when learners think about and process their experiences, allowing them the opportunity to make sense of and derive meaning from their experiences. *Constructivism* is the theory that argues that knowledge is constructed when the learner draws ideas from his/her own experiences and connects them to new ideas. *Cooperative Learning* puts a premium on active learning achieved by working with fellow learners as they all engage in a shared task. *Discovery Learning* and *Inquiry-based Learning* support the idea that learners learn better when they ask relevant questions and make use of personal experiences to discover facts, relationships, and concepts. Given that the K to 12 Reform is still in its first cycle of implementation, these approaches are being emphasized in training for teachers and are also happening in the classrooms.

Teaching materials in mathematics

The use of appropriate devices and tools in teaching mathematics is being emphasized in teacher preparation and continuing professional development programs, including the use of manipulative objects, measuring devices, calculators and computers, smart phones and tablet PCs, and the internet (DepEd, 2013). DepEd has encouraged use of instructional materials with integration of ICT. Smart boards and internet have been set up in many public schools to increase access of students and teachers to more advanced educational technologies.

The DepEd has also its own Learning Resources Management and Development System (LRMDS) designed to support increased distribution and access to learning, teaching and professional development resources at the region, division and school/cluster levels (DepEd, 2016).

The LRMDS provides access to quality resources in the implementation of the whole K to 12 Curriculum. These include:

1. information on the quantity, quality and location of textbooks and supplementary materials, and cultural expertise;
2. learning, teaching and professional development resources in digital format and the location of resources in print format and hard copy; and
3. standards, specifications and guidelines for assessing & evaluating, acquiring & harvesting, modification, development and production of resources.

It is also a quality assurance system providing support to DepEd Regions, Divisions and Schools in the selection and acquisition of quality digital and non-digital resources in response to identified local educational needs.

The DOST has also given its share as to how teachers could best teach mathematics. According to De Leon (2016), the DOST promotes the teaching of mathematics and science through use of interactive materials to enable the Filipino students to be globally competitive. Through its Science Education Institute (SEI), in partnership with the Advanced Science and Technology Institute (ASTI), a set of interactive courseware in teaching science and mathematics has been developed to upgrade and improve science and mathematics education in the country. This courseware is provided for free to public schools and is also made available as online resources for teachers and students in Mathematics (DOST-SEI, 2016).

There are other initiatives to support the learning of Mathematics by Filipino children. One is the Knowledge Channel Foundation, Inc. (KCFI), a non-profit organization founded in 1999, which is dedicated to uplifting the lives of Filipinos from poverty through education. The foundation provides the learners with access to multimedia learning materials through Knowledge Channel (KCh) television, KCh Online and KCh Portable Media Library. It is the first and only curriculum-based transmedia system focused on Philippine basic education. Following DepEd's curriculum for Early Childhood Development, K to 12, and Alternative Learning System (ALS), KCh produces and acquires videos, games, and applications for the learners and their teachers. These learning resources cater to various topics in key subject areas in the K to 12 curriculum, including mathematics (KCFI, 2014).

Recently, KCh developed a new series that focuses on overcoming children's fear of mathematics to pave the way for its better learning. It developed *MathDali*, which are videos starring five actors and actresses, as the foundation's latest contribution to K to 12 Reform. This series aims at providing mastery of mathematical concepts and skills leading to better learning outcomes (KCh, 2016). *Mathdali* is aired under the mathematics curriculum block of KCh. The series revolves around the story of a math enthusiast, who helps his friends' quests for learning mathematics. Each episode illustrates mathematics concepts taught through real-life situations. It also aims to develop conceptual understanding, critical thinking, and problem solving in mathematics in new ways that make learning interesting and exciting.

In this K to 12 Reform in the Philippines, the production of instructional materials for the delivery of the new mathematics curriculum is deemed necessary. Local publishing companies have also responded to DepEd's call to produce textbooks and other instructional materials like workbooks, modules and ICT-based materials to support the implementation of the ongoing reform.

Professional development program for mathematics teachers

As defined in the Philippines' Code of Ethics for Professional Teachers (Resolution No. 435 s. 1997), every teacher shall participate in the Continuing Professional Education (CPE) program of PRC and shall pursue such other studies as will improve their proficiency, enhance the prestige of their profession, and strengthen their virtues and productivity to be nationally and internationally competitive. Through RA 10912, known as the Continuing Professional Development (CPD) Act of 2016, CPD was set as a mandatory requirement in the renewal of professional license and accreditation for the practice of the professions conducted every five years. Given this law, teachers have to continue their professional growth to strengthen their content knowledge and pedagogical skills to engage with the continual developments in education.

In the case of mathematics teachers, there is a framework from which they could anchor their professional growth beyond the minimum required degree and licensure for employment. In 2011, the Science Education Institute (SEI) of DOST, together with MATHTED, Inc., developed a framework for Philippine mathematics teacher education. This framework defines the qualities of effective mathematics teachers in terms of their content knowledge, pedagogical knowledge and management skills. The framework also mapped out four levels for professional growth of mathematics teachers, beginning at novice level and then extending to emerging, accomplished and then expert levels respectively (SEI-DOST & MATHTED, 2011).

In 2016, the Research Center for Teacher Quality (RCTQ), based at PNU but managed collaboratively with the University of New England (UNE) in Armidale, Australia, also developed a set of professional standards for teachers at four career stages from a beginning teacher into a proficient, highly proficient and then a distinguished teacher. This set of standards, called the Philippine Professional Standards for Teachers (PPST) was built on the National Competency-based Teacher Standards (NCBTS) developed in 2007. It is a set of standards anchored on the principle of lifelong learning (RCTQ, 2016).

In a study, conducted by the World Bank Group and Australian Aid, that assesses the content knowledge of teachers including those teaching mathematics, it was reported that an average elementary or high school teacher could answer correctly fewer than half of the questions on the subject content tests (Al-Samarrai, 2016, p. 3). Al-Samarrai even cited that the median mathematics high school teacher was able to provide completely correct answers to only 31 percent of the questions in the test that is aligned with the new K to 12 curriculum in mathematics (p. 3). The results revealed that mathematics teachers need significant improvement in their content knowledge and skills to provide effective instruction in the classroom. The report points out the need for high quality and regular professional development opportunities for teachers. According to Al-Samarrai (2016, p. 7), studies from both developed and developing countries have shown that well-designed in-service training for teachers can increase their content knowledge, improve their methods of instruction, and ultimately improve student learning outcomes.

In response to the need for high quality and regular professional development opportunities for teachers, DepEd Order No. 35, s. 2016 has been formulated to scale up the practice of school-based learning action cells (LAC). In DepEd, LAC is a professional learning community for teachers to help them improve their practice and learner achievement. In LAC, groups of teachers are engaged in collaborative learning sessions to solve challenges encountered in the school. The key aspects of the process are ongoing collaborative learning or problem solving within a shared domain of professional interest, self-directed learning, reflective practice leading to action and self-evaluation, and collective competence (DepEd, 2016, p. 3). Suggested topics for LACs include content and pedagogy of the K to 12 Basic Education Program, which hopefully could address the challenge posed particularly to the mathematics teachers as revealed in the World Bank report.

Moreover, DepEd has also its Teacher Education Council (TEC), that usually takes care of the Teacher Induction Program (TIP) aimed at enhancing the content knowledge and skills of teachers with three years or less in teaching experience. DepEd with TEIs also conduct mass training programs every year to retool teachers in the implementation of the K to 12 curriculum. Schools and universities also initiate their own CPEs for their teachers and staff, usually before the start of classes. The CPE programs usually focus on school development for the improvement of students' academic learning.

There are other institutions and groups that help strengthen teachers' content knowledge and pedagogical skills in teaching Mathematics. For example, the Fund for Assistance to Private Education (FAPE) usually takes care of the CPE of private school teachers. The National Institute for Science and Mathematics Education Development (NISMED), a research unit and an extension arm of the University of the Philippines (UP) established in 1997, helps raise the level of teaching and learning of science and mathematics in the Philippine educational system. Its mission is to be a major source of quality information and services for stakeholders in science and mathematics education. It has produced sourcebooks, posters, improvised equipment, and non-print materials for use of mathematics teachers. It has also popularized the practical work approach to teaching mathematics, exemplified by hands-on, minds-on and hearts-on activities for the learners (UP-NISMED, 2016).

Most of UP-NISMED's programs and projects have national or regional significance. However, one of its current projects is the Collaborative Lesson Research and Development (CLRD) Project, which introduces *Lesson Study* as a model of continuing professional development for teachers and promotes teaching mathematics through problem solving (UP-NISMED, 2016). According to UP-NISMED, the Lesson Study that originated in Japan in 1872 (Makinae, 2010), is a school-based teacher-led continuing professional development model. In this model, teachers work collaboratively to:

1. formulate goals for students learning in long-term development;
2. develop "research" lessons to bring those goals to life;
3. implement the lesson and document observations;
4. discuss the evidence gathered during the lesson implementation and use it to improve the lesson; and
5. teach and study the lesson again to further refine it (UP-NISMED, 2012).

At present, UP-NISMED has been training mathematics teachers in schools across the country on how to implement such an approach for their professional development.

Active membership in professional organizations is another venue for mathematics teachers' professional development. One professional organization recognized for leading mathematics research and education in the country is the Mathematical Society of the Philippines (MSP). MSP is the Philippines' premiere professional organization dedicated to the promotion of mathematics research and education in the country (MSP, 2016). Founded in 1973, it has grown from a small Manila-based group of math educators to a nation-wide network of individuals, with chapters in the three island groups of Luzon, Visayas and Mindanao. The MSP has represented the country in the International Mathematical Union since 1978. It conducts the annual Philippine Mathematical Olympiad and trains the Philippine team to the International Mathematical Olympiad with the support of DOST and other sponsors. It conducts national conventions and promotes publications of research in mathematics in refereed international journals.

The utilization of national and international conferences as venues for teachers' professional development is now well-established in the Philippines. One organization, formed by the science and mathematics classroom teachers and teacher educators trained at the Regional Center in Science and Mathematics (RECSAM) in Penang, Malaysia, that helps provide opportunities for mathematics teachers' professional development is the Philippine Association of RECSAM Grantees (PARG). PARG organizes biennial conferences as a venue for sharing new teaching methods and trends in mathematics (UP-NISMED, 2001, p. 116).

Another professional organization that organizes conferences as venues particularly for the dissemination of their research in mathematics and mathematics education is the Philippine Council of Mathematics Teacher Educators (MATHTED). In October 2016, for example, the MATHTED Inc. celebrated its 20th year as a national professional organization for mathematics teachers that pursues the advancement of mathematics education in the country through a national conference. MATHTED (n.d.) aims to:

1. provide leadership in improving the quality of mathematics teacher education;
2. promote research that will address relevant issues in mathematics education;
3. facilitate the exchange of information about current research work and teaching methods among mathematics teacher educators; and
4. encourage and foster collaboration with members of other recognized mathematical, scientific, and educational institutions.

These forms of mathematics teachers' professional development will continue to evolve as the Philippines finds its way to a quality mathematics education system that is comparable with the best in the international education community.

Conclusions and Implications

The teaching of mathematics and the preparation of teachers that could teach it effectively is indeed a challenge. In a developing country like the Philippines, there is a great demand to show the quality of its education revealed by observable measures. The consistently dismal performance in mathematics of primary and secondary school students in national and international examinations taken prior to the implementation of the K to 12 Reform, this country's major educational reform, is expected to improve significantly in the future. The increase in the number of years of basic education from 10 to 13 to include compulsory kindergarten and senior high school programs should make the Philippines comparable with the quality of education of high performing countries in the region. The different types of national assessments of learners, including those who would become mathematics teachers, as well as the directions of participation in international assessments in mathematics should enable the Philippines to raise its image in the global education community. At all levels of education, the inclusion of mathematics in examinations, be it national or international, is indicative of how essential mathematical knowledge and skills are. Hence, the current educational reform in the Philippines and the support from various groups of professional organizations and stakeholders of mathematics education in the country give hope towards making the country's education at least comparable with its neighboring countries.

To make the K to 12 Reform work in improving the quality of mathematics education in the Philippines, there are several areas that should be reviewed continually. The curricula of mathematics and mathematics teacher education should be evaluated regularly and benchmarked with internationally accepted standards. The criterion of mastery that has been targeted in national examinations should be re-examined as well, to support proper interpretation and appropriate action on the results. The development of positive attitudes and aptitudes in mathematics among students should also be targeted towards leading them into taking a career aligned with the discipline. The nature of mathematics, which is a language, an art, a science and a tool in everyday life should also give teachers a clear view and enable them to reflect critically on how it should be taught. Teaching practice that is informed by the theories of constructivism and contextualized learning should pave the way to the generation of new approaches and the development of localized instructional materials appropriate to 21st century learners. All efforts from government agencies, organizations and educational institutions should be consolidated and directed to help students gain better learning outcomes, particularly in mathematics, as it is with the increase of educational outcomes that the country's economy is expected to improve as well.

Likewise, the continuing professional development of teachers should be given equal attention too. The studies reported that teachers' content knowledge and pedagogy reflect what they could also provide to their students. Hence, improving their capacity to teach their discipline effectively is expected to also have a positive impact upon their students' learning outcomes. The use of more effective and efficient professional development programs should find its way to all schools of whatever level and typology in all regions in the Philippines. Future studies could also be conducted to continuously document the reform movement of the way mathematics is taught to the Filipino people.

References

Al-Samarrai, S. (2016.) *Developing a proficient and motivated teacher workforce in the Philippines*. Philippines education note; no. 3. Washington, D.C.: World Bank Group.

Benito, N.V. (2010). *National achievement test: an overview*. Retrieved from https://www.affordable-learning.com/content/dam/corporate/global/palf/pdf_files/ National_Achievement_Test_Dr%20Benito.pdf

Commission on Higher Education (2012). CMO No. 46, s. 2012 *"Policy-standard to enhance quality assurance (QA) in Philippine higher education through an outcomes-based and typology-based QA"*. Quezon City, Philippines: CHED

Commission on Higher Education (2004). *CHED Memorandum No. 30. S. 2004 "Revised Policies and standards for undergraduate teacher education program"*. Quezon City, Philippines: CHED

De Leon, E. A. (2016). *Use interactive materials to teach science and math, expert advises in DOST scientific conference*. Retrieved from www. dost.gov.ph

Department of Education (2016). *DepEd Order. No. 35, s. 2016 "The learning action cell as a K to 12 basic education program school-based continuing professional development strategy for the improvement of teaching and learning"*. Retrieved from www.deped.gov.ph

Department of Education (2016). *DepEd Order. No.55, s. 2016 "Policy guidelines on the national assessment of student learning for the K to 12 Basic Education Program"*. Retrieved from www.deped.gov.ph

Department of Education (2016). *Learning Resources Management and Development System (LRMDS)*. Retrieved from http://lrmds.deped.gov.ph/terms on November 26, 2016.

Department of Education (2015). *DepEd Order No. 57. s. 2015: Utilisation of the Early Grade Reading Assessment (EGRA) and Early Grade Math Assessment (EGMA) tools for system assessment*. Retrieved from www.deped.gov.ph

Department of Education (2015). *Draft Plan of the Policy Guidelines on system assessment in the K to 12 basic education program. Discussed during the meeting of the National Technical Working Group of K to 12 Assessment held on January 26, 2015*.

Department of Education (2015). *Technical Working Group on the system assessment of the K to 12 Reform*. Retrieved from www.deped.gov.ph

Department of Education (2013). *Fact Sheets*. Retrieved from www.deped.gov.ph

Department of Education (2013). *K to 12 Curriculum Guide in Mathematics, Grades 1 to 10*. Pasig City, Philippines: Department of Education.

Department of Science and Technology (DOST) Science Education Institute (SEI) (2016). *Development of interactive science and mathematics courseware for secondary-level schools*. Retrieved from www.sei.dost.gov.ph

Gonzales, P., Guzmán, J.C., Partelow, L., Pahlke, E., Jocelyn, L., Kastberg, D. and Williams, T. (2004). *Highlights from the Trends in International Mathematics and Science Study (TIMSS) 2003 (NCES 2005–005)*. U.S. Department of Education, National Center for Education Statistics. Washington, DC: U.S. Government Printing Office. Retrieved from http://nces.ed.gov/pubs2005/2005005.pdf

Knowledge Channel Foundation, Inc. (2014). *About the Knowledge Channel Foundation*. Retrieved from http://kchonline.ph

Knowledge Channel Foundation, Inc. (2016). *MathDali. changing mindsets, championing math culture*. Retrieved from http://kchonline.ph

Makinae, N. (2010). *The Origin of Lesson Study in Japan*. Retrieved from http://www.lessonstudygroup.net/lg/readings/TheOriginofLessonStudyinJapanMakinaeN/TheOriginofLessonStudyinJapanMakinaeN.pdf

Manapat,R. (2011). *Mathematical ideas in early Philippine society*. In Philippine Studies Vol. 59, no. 3. Quezon City, Philippines: Ateneo de Manila pp. 291–336.

Mathematics Society of the Philippines (2016). *About MSP*. Retrieved from http://mathsociety.ph/index.php?option=com_content&view=article&id=7&Itemid=108

Mathematics Teachers Association of the Philippines (2016). *DepEd Advisory No. 175. S.2016. MTAP DepEd saturday programs in mathematics for regular and talented learners*. Retrieved from www.deped.gov.ph

Mathematics Trainers' Guild, Philippines (2016). *MTG Profile*. Retrieved from www.mtgphil.org

Metrobank-MTAP-DepED Math Challenge, MMC, (2016). *About Us*. Retrieved from https://www.facebook.com/pg/MetrobankMathChallenge/about/?ref=page_internal

Ogena,E.O., Brawner, F.G., and Ibe, M.D. (2013). Preparing teachers of Mathematics in the Philippines. In J. Schwille, L. Ingvarson, & R. Holdgreve-Resendez (eds), *TEDS-M Encyclopedia, a guide to teacher education context, structure and quality assurance in 17 countries, findings from the IEAteacher education and development study in Mathematics*. Amsterdam, Netherlands: International Association for the Evaluation of Educational Achievement (IEA)

Ogena, E.B., Laña, R.D., and Sasota, R.S (2010). *Performance of Philippine high schools with special curriculum. In the 2008 Trends in International Mathematics and Science Study (TIMSS-Advanced)*. Retrieved from www.nap.psa.gov.ph

Philippine Council of Mathematics Teachers (MATHED), Inc (n.d.). *Vision*. Retrieved from http://mathted.weebly.com

Philippine National Research Center for Teacher Quality (2016). *Philippine professional standards for teachers*. Manila: Philippine Normal University

UP National Institute of Science and Mathematics Education Development and the Foundation for the Advancement of Science Education, Inc. (2001). *One hundred years of science and mathematics education in the Philippines*. Quezon City, Philippines: UPNISMED

Republic of the Philippines (1973). *Presidential Decree No. 146, Upgrading the quality of education in the Philippines by requiring all high school graduates seeking admission to post-secondary degree programs necessitating minimum of four year's study to pass a National Entrance Examination and appropriate funds therefor*. Retrieved from http://www.gov.ph/1973/03/09/presidential-decree-no-146-s-1973/

SEI-DOST & MATHTED, (2011). *Framework for Philippine mathematics teacher education*. Manila: SEI-DOST & MATHTED.

Tatto, M.T. et al. (2012). *Policy, Practice, and Readiness to Teach Primary and Secondary Mathematics in 17 Countries*. Netherlands: International Association for the Evaluation of Educational Achievement (IEA).

University of the Philippines National Institute for Science and Mathematics Education Development (2016). *About Us*. Retrieved from http://www.nismed.upd.edu.ph/about-nismed/

University of the Philippines National Institute for Science and Mathematics Education Development, UP-NISMED, (2012). *What is lesson study?*. Retrieved from http://lessonstudy.nismed.upd.edu.ph/2012/07/lesson-study.html

Wikipedia (2016). *Education in the Philippines during Spanish rule*. Retrieved from https://en.wikipedia.org/wiki/Education_in_the_Philippines_during_Spanish_rule

Wikipedia (2016). *Education in the Philippines during the American rule*. Retrieved from https://en.wikipedia.org/wiki/Education_in_the_Philippines_during_the_American_rule

Wikipedia (2016). *History of the Philippines before 1521*. Retrieved from https://en.wikipedia.org/wiki/History_of_the_Philippines_(before_1521)

About the Authors

Ester B. Ogena, Ph.D. earned her bachelor's degree in Mathematics at the Philippine Normal University (PNU) and her master's degree in Mathematics Education and doctorate degree in Educational Research and Evaluation at the University of the Philippines. She holds postgraduate certificates, among others, on operations research from international Center for Pure and Applied Mathematics at Grenoble University, France and on science and mathematics education from the International Science Education Center in Kiel University, Germany. She is currently the University President of PNU but prior to her assignment to this University, she was Director of the Science Education Institute of the Department of Education and Technology of the Philippines. She has initiated important pioneering programs on science and mathematics education in the country and likewise for the university. She is the founding chairperson of the ASEAN Teacher Education Network (ASTEN) which is actively pursuing initiatives for improving the quality of teacher education in the region.

Marilyn U. Balagtas, Ph.D. earned her doctorate in Educational Research and Evaluation at the University of the Philippines. She took her bachelor's degree in Elementary Education with specialization in Mathematics and her master's degree in Mathematics Education at the Philippine Normal University. She has been trained in professional education at Queensland University of Technology, Australia and on the Development and Evaluation of Problem Solving Skills in Mathematics at SEAMEO-RECSAM, Penang, Malaysia. She was the Inaugural Director of the Philippine National Research Center for Teacher Quality (RCTQ) funded by the Australian Government to support the K to 12 Reform of the Philippines. She is a University Professor and currently the Dean of the College of Flexible Learning and ePNU. She is also the President of the Philippine Association of Teachers of Educational Foundations-United Professionals for the Development and Advancement of Teacher Education, Inc (PATEF-UPDATE) and Board Member of the Philippine Educational Measurement and Evaluation Association.

Rosemarievic V. Diaz, Ph.D. obtained her Ph.D. in Science Education (major in Math) at the De La Salle University where she graduated with distinction and as a DOST-SEI Scholar. She took Master of Science in Mathematics at the same university and obtained Bachelor of Science in Mathematics at the Philippine Normal College as magna cum laude and as a DOST-SEI Scholar. She also attended Professional Studies in Education at Queensland University of Technology, Brisbane, Australia under the RP-Australian Project in Basic Education (PROBE). She is currently a Full Professor at the College of Flexible Learning and ePNU but prior to her assignment to this College, she served as the Vice President for University Relations of PNU. She is the current president of the Philippine Council of Mathematics Teacher Educators, Inc. (MATHTED), a professional organization of mathematics teacher educators.

Chapter VI

INDONESIA: History and Perspective on Mathematics Education

Sitti Maesuri Patahuddin
University of Canberra, Australia

Stephanus Suwarsono
Sanata Dharma University, Yogyakarta, Indonesia

Rahmah Johar
University of Syiah Kuala, Aceh, Indonesia

Introduction

Mathematics education has a very critical role in the development of Indonesian education and society. To begin, we provide an overview of Indonesian geography and demography followed by the story structured within three main consecutive periods: Pre-colonial era, Colonial era (1600–1945) and Post-Independence Era (1945 till now). We will end the chapter by highlighting some challenges and provide possible solutions.

Geography and Demography

More than 250 million people inhabit Indonesia thus making it the world's fourth most populated nation, after China, India and the United States. Approximately 60% of the population lives in Java whose area is about 10% of the whole land area of Indonesia. Population density varies from 15,173 people per square kilometre in Jakarta (Indonesian capital city) to 9 in West Papua and 8 in North Kalimantan (Badan Pusat Statistik, 2016).

Indonesia consists of around 17,500 islands along the equator between Asia and Australia and around 35% of these islands are inhabited. The major islands of Indonesia are Java, Sumatra, Kalimantan and Papua. Indonesia is also recognised as highly diverse ethnically and linguistically with approximately 600 languages or dialects spoken across islands and around 3,000 ethnic groups. Therefore, the official motto Bhinneka Tunggal Ika (Unity in Diversity) and the country's official language (Bahasa Indonesia) are very critical. Indonesia is also recognised as the world's largest population of Muslims consisting of almost 90% of the population. Other official religions included Christian, Catholic, Hindu, and Buddhist (Alwasilah & Furqon, 2010). This geographic and demographic situation provides certain challenges in the development of mathematics education in Indonesia.

The Pre-Colonial and Colonial Period (1600–1945)

Indonesia has undergone colonization by a number of European nations (Portugal, the Netherlands, Britain), and then occupied by Japan during the Second World War. Prior to colonization, Indonesia was inhabited by people from other parts of the world, who brought with them their own cultures. Therefore, it is appropriate to differentiate the times according to the people who have inhabited or controlled the country.

- The Pre-Colonial Era
- The Portuguese Era (until 1605)
- The First Dutch Era (1605–1811)
- The British Era (1811–1816)
- The Second Dutch Era (1816–1891, the Liberal Era)
- The Third Dutch Era (1892–1942)
- The Japanese Occupation (1942–1945)
- The Indonesian Era (Post–Independence Era, 1945 till now)

Pre-Colonial Era

According to archaeological findings, the Indonesian archipelago began to be inhabited by humans more than 500,000 years ago by a dark-skinned race of people. Later many migrants came from parts of mainland of Asia,

Vietnam and Cambodia. These new migrants far outnumbered the earlier dwellers (Marsaban, 1968).

Marsaban also stated that, around the turn of the first century AD, a different kind of migrant came to Indonesia. These were the Hindu traders who came in great numbers, bringing with them a highly developed religion and culture, which soon deeply influenced the way of life and customs of original inhabitants. Hinduism had a glorious period culturally and artistically for over 1400 years. Its influence on Indonesia's culture can still be felt today. Traders from India also brought Buddhism to Indonesia.

During the seventh century, a great Indonesian Kingdom rose gloriously in Sumatra. Sriwijaya would eventually become a great commercial and naval power. Its influence extended to Malaya (Malaysia), the Philippines, parts of Indo-China, Cambodia, and Southern China. Marsaban (1968) and Pranoto (2016) stated that at that time Sriwijaya was a great centre for learning, on a par with Nalanda, in India. Pranoto also quoted Yijing (sometimes also spelled I Ching), a Chinese scholar, who had studied both in Sriwijaya and Nalanda, who said that the curriculum and the depth of knowledge studied in Sriwijaya was equal to that which is studied in Nalanda. Yijing also wrote (about 1300 years ago) that the knowledge studied in Nalanda was not just theology, but other forms of science like mathematics, logic, astronomy, and medicine as well as language and art.

However, according to Pranoto (2016), although there were references to the era of Sriwijaya education being quite advanced, there is no written evidence that can be used as proof of the great academic achievement from the Sriwijaya era. Quoting other experts, Pranoto said that the proof may not lie in the form of a written legacy, but in the artefacts that remain such as temples, art forms (dance, music and artworks), ideas and beliefs that still exist today. For example, Pranoto argued the process of building such a complex structure as Borobudur temple in Central Java could not be done without a deep knowledge and application of mathematics and physics.

Another strong kingdom emerged in Indonesia during the thirteenth century. The Majapahit Empire, (a Hindu empire), was established in East Java. This sovereignty soon spread throughout the entire archipelago (Marsaban, 1968).

The Majapahit Empire reigned until 1487. Marsaban also stated that during the era of Hindu Kingdoms, many buildings called *"padepokan"* were built around the temples. A *"padepokan"* was a place where people often gathered to discuss community problems. It was also used as a learning centre for the youth, where religious teachings were delivered.

Although Hinduism has been adopted only by a minority in the Indonesian population, the 1400 years of Hindu kingdoms have left abiding marks on Indonesian culture that are still present today, in the forms of architectural heritage (e.g., Prambanan temple in Yogyakarta) and dramatic heritage (puppet performances, traditional dances and music) (Marsaban, 1968).

Towards the end of this era the Islamic influence was on the rise. Traders and their followers from Arab countries, India, and Persia introduced the Moslem faith to Indonesia. At that time many Moslem centres of learning, called "Madrasah" or "Pesantren" were founded and these centres were utilised to teach the religion of Islam.

Portuguese Era (until 1605)

The Portuguese were the first European traders to come to Indonesia, arriving in the early sixteenth century (Marsaban, 1968). They colonised the eastern parts of Indonesia to gain control over the spice trade (Nasution, 1983). Nasution also mentioned that the Portuguese traders arrived in Indonesia accompanied by missionaries, who introduced Catholicism. As well as introducing their religion (Catholicism), the Portuguese missionaries also built schools in the modern sense of the word. In these schools, religion (Catholicism) was the main subject taught. However, unlike the "padepokan" during the Hindu time or "pesantren" established by the Moslem people earlier, these schools also taught reading, writing using Roman script and arithmetic. Quite possibly this is the first time that mathematics was taught in an Indonesian school (in the modern sense of the word). Nasution (1983) noted that during the Portuguese rule they successfully spread the use of the Portuguese language among the Indonesian people and the popularity and use of Portuguese was equal to that of Melayu (the origin of Bahasa Indonesia). Portuguese dominance was weakened during the war against the Indonesian sultanates, and their rule was ended by the Dutch forces in 1605.

The First Dutch Era (1605–1811)

A common Indonesian viewpoint is that the Dutch colonised Indonesia for three and a half centuries. This is true for only some parts of this giant archipelago.

The presence of the Dutch colonial power in Indonesia initially was through the Dutch East India Company *(Verenigde Oost-Indische Compagnie,* VOC). Its activities were conducted for commercial purposes, namely that of monopolising exports (e.g., spice) and enforcing on the inhabitants a new agricultural pattern designed to serve Dutch commercial interests (Marsaban, 1968).

As the Dutch were Protestants (Calvinists) they soon stopped the spread of Catholicism and promoted the Protestant faith. They (VOC) also established schools in the eastern parts of Indonesia, the first school being founded in Ambon in 1607 for Indonesian children (Nasution, 1983). Soon the number of schools in Ambon rapidly increased. Then in 1630 they began to establish schools in Batavia (the former name for Jakarta). The main purposes for establishing these schools was to educate Dutch and Indonesian children to become competent workers for the VOC, and to spread the Protestant form of Christianity. At that time the duties of teachers in these schools were to teach children about Christianity (the Protestant faith). These included prayer, singing, attending church and obedience to parents, rulers (the government and its representatives), and teachers, as well as reading and writing (Nasution, 1983).

By the middle of the eighteenth century, education in Indonesia had undergone a serious decline. The education provided by the VOC in Indonesia was regarded as a total failure in the eyes of the Dutch Government. It has been stated that Jakarta had a population of 16,000 school aged children) but only 270 of these attended school. In Surabaya only 24 attended school, and throughout the whole of Java there were only 350 attendees (Nasution, 1983, p. 7). One explanation for the lack of attendance, as viewed by the Dutch Government of the time, was the failure of the VOC to use the Dutch language in schools. It was very difficult to find teachers who could speak Dutch. Moreover, Portuguese and Bahasa Melayu (the origin of the Indonesian Language) were still the most common languages.

In 1799, the VOC went bankrupt and was liquidated. Its position in Indonesia was taken over by the Dutch Government (Marsaban, 1968; Nasution, 1983), until the British began their rule.

The British Era (1811–1816)

British forces ruled Indonesia from 1811 to 1816. Due to the occupation of The Netherlands by Napoleon Bonaparte (of France), in 1811 the British seized the opportunity to occupy (Java) Indonesia (Marsaban, 1968, p. 114). However, according to Nasution (1983, p. 7), the presence of the British in Indonesia did not bring any changes to the field of education.

The Second Dutch Era (1816–1891, also called the Liberal Era)

After the defeat of Napoleon Bonaparte, the Netherlands regained their independence and Indonesian rule was returned to the Dutch in 1816 (Nasution, 1983, p. 7). During this time the Government of the Netherlands was controlled by people who held liberal views, about how education could be used as a vehicle for economic and social progress. Therefore, it was considered that the Dutch had a responsibility to develop an education system both for Indonesian children as well as Dutch children who lived in Indonesia. They believed that providing a better education system for Indonesian children would yield a twofold benefit. It would be returning something good to Indonesia (as a colonised country which had provided much profit for the Netherlands), and it would also benefit the Netherlands, as educated Indonesian people would be able to better serve the Dutch administration in Indonesia. (This view was also called the Ethical Policy).

Establishing ELS and HBS education

Initially however Dutch officials only established one primary school, which was located in Jakarta and was exclusively for Dutch children. It was called ELS *(Europese Lagere School,* European Lower School) and was a seven-year program delivered in the Dutch language. In 1817, and then for some years after, they established more ELS schools in other areas of Java (Nasution, 1983, p. 9). These schools began to accept Indonesian students from wealthy or high-ranking families *(priyayi).* However, the number of such children was very small. In 1867, the first secondary school called HBS *(Hogere Burger School),* was opened in Jakarta for Dutch students. HBS provided a five-year program where graduating Dutch students could continue their education at a university in the Netherlands. Several years later more HBS's were opened in Surabaya and Semarang.

Establishing schools for Indonesian students

In 1848 the Dutch officials began to establish primary schools for Indonesian students (Nasution, 1983). These schools taught reading, writing, language (local language, or Bahasa Melayu), and arithmetic.

The Third Dutch Era (1892–1942)

In 1892, primary schools for Indonesian children (the native children) were reorganised. This was because authorities believed that the present circumstance did not provide a suitable education for those students who would be recruited to work for the Dutch and provided too much knowledge for those who would not be recruited (Nasution, 1983, p. 50). On the basis of this view, "primary schools" for Indonesian children were differentiated into two types. First Class Schools *(Sekolah Angka Siji, Eerste Klasse School),* designed for the Indonesian children who came from wealthy high ranking families, and Second Class Schools *(Sekolah Angka Loro, Twede Klasse School)* for the Indonesian children who came from ordinary or lower class families.

The curriculum for these primary schools consisted of reading and writing in the local language (using characters and Latin characters), reading and writing using Melayu (the origin of Bahasa Indonesia), arithmetic, Indonesian geography, physics, the history of the island where the students lived, drawing, and land measurement. Singing could be taught but it was not compulsory (Nasution, 1983, p. 52).

At a certain stage (in 1914) the First Class School was developed (extended) to become HIS *(Hollands Inlandse School)* in order to increase the quality of the school in the same standing as ELS and HCS *(Hollands Chinese School,* a primary school designed especially for Chinese children, that used Dutch as the medium of instruction and was regarded as having the same standing as ELS which was set up in 1908) (Nasution, 1983, p. 108, 114).

In 1903, another type of school was opened in Bandung and Jogjakarta (now called Yogyakarta), which was named MULO *(Meer Uitgebreid Lager Onderwijs)* and it needed four years to complete. MULO was meant to be a type of primary school with an extended program, as a continuation of HIS. MULO was not regarded as a secondary school, although MULO could only be entered by a student who had graduated from HIS. MULO was regarded as a mediating school between primary school and the secondary school that had existed at that time, namely HBS, due to the fact that many children who went to HBS directly from primary school dropped out before they arrived at the final year of HBS (year 5) (Nasution, 1983, pp. 122–123). The level of complexity of the education in MULO was much lower than that in HBS, therefore many students who did not feel very capable academically but aspired to enter HBS could go to MULO first, and when they graduated from MULO they could enter grade 4 in HBS. Having graduated from HBS, they could enter university education in the Netherlands (for those who could afford it).

HBS in Indonesia had the same curriculum as that in HBS in the Netherlands (Nasution, 1983, p. 131). Its academic quality was also very high (p. 133). The teachers at HBS had to have a Ph.D. degree or its equivalent (p. 134). At that time, a Ph.D. degree had not yet been obtainable in Indonesia, as no tertiary learning had been established. However, a Ph.D. degree could be obtained in the Netherlands. As a result, teachers at HBS were mainly Dutch people.

In Tables 1 and 2 below, the subjects taught at MULO and those taught at HBS are presented.

Although MULO was designed as an extension of HIS to prepare its graduates for HBS, the relationship between the two institutions was not satisfactory and some experts in the field of education had objections about MULO as preparation for HBS. As a result, another model of secondary school was established. Its name was *Algemene Middelbare School* (AMS), and the first AMS was opened in 1919. A diploma from AMS was officially regarded as having the same standing as that of HBS. AMS graduates, like their counterparts from HBS, could now directly enter a university in the Netherlands or Indonesia (when higher education became available). Therefore, the founding of AMS also stimulated the need for the establishment of tertiary learning institutions in Indonesia.

The AMS curriculum consisted of two academic streams, namely Stream A and Stream B. Stream A students studied literature and history, while Stream B students concentrated on mathematics and physics. Stream A was also subdivided into two lines, namely A-I (Line 1) and A-II

Table 1. Subjects taught at MULO and the number of periods per week (Nasution, 1983, p. 124).

Subjects	Number of periods per week in each grade		
	I	II	III
Reading	3	3	2
The Dutch Language	5	4	4
Writing (Occasional)	Occasional	Occasional	Occasional
Arithmetic and Mathematics	8	9	7
Dutch and Colonial History	1	1	2
World History	1	1	1
Geography	3	3	3
Physics	3	3	4
French Language	2	4	4
English Language	4	4	3
German Language	4	3	4
Drawing	2	2	2
Total number of periods per week	36	36	36

Table 2. Subjects taught at HBS and the number of periods per week (Nasution, 1983, p 124).

Subjects	Number of periods per week in each grade				
	I	II	III	IV	V
Arithmetic and Algebra	5	5	3	2	1
Mathematics	4	4	4	4	4
Mechanics	-	-	-	3	2
Physics	-	-	4	4	3
Chemistry	-	-	2	4	5
Botany	1	1	1	1	1
Biology	1	1	1	1	1
Cosmography	-	-	-	1	1
State Laws and Regulations	-	-	1	1	1
Accounting	1	-	1	1	1
History	3	3	3	3	3
Geography	3	3	2	2	1
Dutch Language	5	4	4	3	3
French Language	4	4	4	3	3
German Language	4	4	4	3	3
English Language	4	4	4	3	3
Hand Drawing	2	3	2	2	2
Line Drawing	-	-	2	2	1
Total number of periods per week	36	36	43	43	40

(Line 2). A-I concentrated on the study of Eastern Classical Literature, while A-II concentrated on Western Classical Literature (Nasution, 1983, p. 138). Core subjects were taught in both lines of A-I and A-II which all students completed. These subjects were Dutch, Melayu, English, history, geography, state acts and regulations, mathematics, botany, zoology, and physical education. The specialised subjects for A-I (Eastern Classical Literature) were Javanese, archaeology, Indonesian ethnology, physics, chemistry, hand drawing, German, and French as an elective. The specialised subjects for A-II (Western Classical Literature) were the same as those taught in AI, except that Latin replaced the Javanese, and the extra subject of accounting *(tata buku)* was also taught. The subjects for

Stream B were physics, chemistry, mathematics, cosmography, line drawing, and German. French was also available as an elective. The course of study for AMS was 3 years after MULO (Nasution, 1983, p. 102). The high schools (SMA in Bahasa Indonesia) that exist today in Indonesia have been based on the AMS model.

The academic quality of Indonesian children

As has been stated earlier, although ELS was specifically designed for Dutch children, some Indonesian children who came from rich or high ranking Indonesian families were also accepted as students. Regarding the academic abilities of those Indonesian children, Nasution (1983, p. 105) wrote:

> *Guru-guru Belanda mengakui kemampuan anak-anak Indonesia dalam segala mata pelajaran, sekalipun semua pelajaran diselenggarakan dalam Bahasa Belanda* ("The Dutch teachers acknowledged the abilities of Indonesian children in all school subjects, although those subjects were taught in Dutch language").

Further, Nasution wrote:

> *Prestasi akademis anak Indonesia tidak kalah dari anak-anak Belanda seperti nyata dari persentase lulusan masuk HBS atau ujian pegawai rendah.* ("The academic achievement of Indonesian children was not lower than that of the Dutch children, as evidenced from the percentage of graduates accepted at HBS or graduates passing the test for lower government officials").

Some examples of the mathematical problems given to children

In order to give an insight into the mathematical problems given to Indonesian students during the colonial era, the following two examples were found in the final examination of Kweek School. (Kweek School was a school at the level of junior high school, that prepared students to become teachers at the primary school level) (Nasution, 1983, pp. 43–44).

Aritmetika: Seseorang membeli 100 gelas seharga f 50,- dan dijualnya dengan harga f 0,80 per buah. Oleh sebab ada sejumlah gelas yang pecah, ia hanya memperoleh 13/15 dari laba yang diharapkannya. Berapa banyak gelas yang pecah? (Kweek School, Padang Sidempuan, Sumatra, Indonesia, 1891).

Arithmetic: A buyer bought 100 glasses for f 50,- and he sold the glasses for f 0,80 each. Because some of those glasses were then broken, the profit that he obtained was only 13/15 of the profit that he had expected. How many were broken? (Kweek School, Padang Sidempuan, Sumatra, Indonesia, 1891).

Geometri: Panjang sebuah prisma segiempat tiga kali dan lebarnya setengah dari suatu kubus. Bila luas prisma itu 50m²lebih dari kubus itu, berapakah isi prisma itu? (Kweek School, Bandung, West Java, Indonesia, 1891).

Geometry: The length of the base of a rectangular prism is three times of the length of an edge of a cube, and its width is a half of the length of the edge. If the total area of the faces of the prism is 50 cm² bigger than that of the cube, what is the volume of the prism? (Kweek School, Bandung, West Java, Indonesia, 1891).

It can be seen that the above problems were not so simple for students who were in their fourth year of junior high school (equivalent with Year 10 of students in Australia). Nasution (1983) reported that Indonesian students' success rate for these types of mathematical problems was approximately 90%.

The establishment of a tertiary learning institution in Indonesia

The first tertiary learning institution was established in Indonesia in 1920. After the First World War, the relationship between Indonesia and the Netherlands was interrupted, and Indonesian students found it very difficult to go to the Netherlands to study at a university level. Therefore, many local students needed the opportunity to attend tertiary education and the first institution was established in Bandung. It was called Technische Hogeschool (later to be renamed Bandung Institute of Technology) and engineering studies were offered. The fact that the first tertiary learning institution of Indonesia was in the field of engineering indicates that Indonesian people had great interest in mathematics and other exact sciences. One of the first Indonesian graduates was Soekarno, who later became the first president of The Republic of Indonesia, in 1945 (Nasution, 1983, p. 144). With the establishment of *Technische Hogeschool* in Bandung, the education system in Indonesia was now complete. Indonesian students could now begin school at the primary level and complete their study in Indonesia with a recognised university degree.

Japanese occupation (1942–1945)

Japanese occupation was only for a short period of time, namely 3.5 years. However, they had a significant influence on the field of education. According to Marsaban (1968), the main legacy the Japanese administration managed to make was the compulsory use of the Indonesian language (Bahasa Indonesia) in schools from the primary level to the tertiary level. Indonesian people finally had the freedom to use their own language.

Post-Independence Era, 1945 Till Now

The consolidation of independence (1946–1965)

The proclamation of Indonesian independence on 17 August 1945 by two Indonesian leaders Soekarno and Hatta marked a new era for Indonesia, including a new era in education. Under the Indonesian Constitution of 1945, every Indonesian was entitled to be educated and the Indonesian Government was obliged to provide one system of education. One act that was designed to implement these rights and obligations was the Educational Act of 1950 which stated that the broad aim of national education was to create democratic citizens who are responsible for the well-being and prosperity of society and the fatherland (Marsaban, 1968).

So, it appeared that the Indonesian Government was trying to set up an education system that was more suitable to Indonesia's needs. However, by looking at the textbooks of mathematics (and of other 'universal' subjects like physics and chemistry) of that time, it appears that many practices from the pre-independent era, or at least from the Dutch colonization period, continued to be used as models. For example, the system of education at SMA (senior high school) replicated the AMS model. Problems that were posed in final examinations in mathematics were similar to those given in the examinations of AMS and HBS. For example, a mathematics textbook for Senior High Schools titled *"Ilmu Ukur Ruang, Jilid II"* ("Solid Geometry, Volume II"), written by Sunarno (1951), published in 1951, used many problems as exercises that were from the final examinations of AMS and HBS many years earlier. Many other mathematics textbooks, and books on other subjects, were written by Dutch writers and were translated into Bahasa Indonesia. For example, a textbook on Algebra and a textbook of Trigonometry, written by CJ Alders, were used widely in the senior high schools at that time.

The situations in junior high and primary schools were similar to that in senior high schools. Namely, the influence of the education system under the Dutch era was still very strong. Three different curricula were implemented during this Independence Consolidation Era: Curriculum 1947, 1952, and 1964 (Mailizar, Alafaleq, & Fan, 2014). These curricula were not significantly different from those prior to 1947. The only change was that students were organised in their first year of senior high into 3 streams: Language and Culture Stream A *(Bagian A)*, Science and Mathematics Stream B *(Bagian B)*, and Economics and Social Science Stream C *(Bagian C)*.

New Order (1966–1998)

Mathematics Education during the "New Order" Era (1966–1998), including Curriculum 1975

As mentioned above, curricula from 1947 till 1968 did not differ very much from each other as far as mathematics education is concerned.

Mathematics content and teaching methods remained the same. The approaches or methods that were widely used by teachers were in the form of a lecture followed by examples displayed on the board and then students were given written exercises to practise the concept. In 1975, when the Indonesian Government introduced Curriculum 1975, big changes (some people called it a 'revolution') were introduced to the mathematics curriculum. These changes were to both the content and teaching methods of mathematics. This new mathematics curriculum was called the Modern Mathematics Curriculum. So 1975 was the boundary where mathematics teaching that used 'traditional' methods and materials ceased and mathematics teaching that used 'new' or 'modern' teaching methods and materials began.

The comparison between the earlier curricula (Curricula Pre-1975) and Curriculum 1975 can be summarised in the following description.

Curricula Pre-1975:

1. Mathematics was still called *Ilmu Pasti* (the "Exact Science"), which was the translation of the Dutch word *"Wiskunde"*.
2. Mathematics taught at primary school was specifically arithmetic.
3. Mathematics taught at junior high school consisted of two parts, namely algebra and geometry, and each part was regarded as two different school subjects. Geometry taught was just plane geometry.
4. Mathematics taught at senior high school consisted of several parts that were taught separately. These were algebra, trigonometry, solid geometry, and analytic geometry.
5. The content of mathematics taught at primary school, junior high school, and senior high school were all "old" or "traditional" topics as mentioned above, which had not included "new" topics like logic, sets, probability, and statistics (Wirasto, 1972).
6. Methods of teaching were generally the lecture model.
7. Theories that were relevant for learning were not discussed or given sufficient attention, let alone used when teaching mathematics.

Curriculum 1975:

1. The subject name for mathematics was no longer *Ilmu Pasti (Exact Science)* but became *Matematika* (which is the direct translation of the word "mathematics").
2. Mathematics taught in primary school now was not just arithmetic, but also included other topics such as sets, equations, inequations, geometry (metric and non-metric geometry, coordinate geometry, and transformation geometry).
3. Mathematics taught in junior high school now consisted of algebra (including sets), arithmetic, geometry (two-dimensional geometry, trigonometry, three-dimensional geometry, analytic geometry, and transformation geometry), probability, and statistics.
4. Mathematics taught in senior high school now consisted of algebra, arithmetic, geometry (two-dimensional and three-dimensional geometry, analytic geometry, transformation geometry, and vector geometry), trigonometry, probability and statistics, logic, and calculus.
5. Mathematics content now included many "new" or "modern" topics such as sets, relations and functions, matrices, vectors, mathematical logic, and many new topics.
6. Sets were being used as foundations for many other topics.
7. Geometry that was now being taught consisted of many variations compared to the "old" geometry.
8. Theories about learning from Piaget, Dienes, Bruner, Gagne, Ausubel and others that were relevant for mathematics teaching were being discussed at mathematics seminars.
9. It was suggested that approaches and methods used in the teaching of mathematics should have more variations (not just lecture method), and efforts should be made to engage students in the process of learning mathematics.

Reformation to the Present (1999–Present)

School system and diversity

The Indonesian school system is huge and diverse. It is the third largest education system in Asia and the fourth largest in the world after China, India and the United States with over 50 million students and 2.6 million teachers in more than 250,000 schools. Indonesia has an educational system administered equally by the Ministry of National Education (MONE) and the Ministry of Religious Affairs (MORA). However higher education has recently been moved under the administration of a newly formed ministry called the Ministry of Research, Technology, and Higher Education. In 1994, a system of 9-years Basic Education was declared. This means that students must complete 9 years of compulsory education—that is 6 years in primary school and 3 years in junior high school (Yeom, Acedo, Utomo, & Yeom, 2002). Furthermore, three-years senior secondary education comprises of 2 streams: general and technical/vocational (Alwasilah and Furqon, 2010).

Mathematics curricula

The national curriculum of Indonesia has been revised multiple times since Independence Day in 1945. Between 1945 and 2016, seven different curricula have been implemented. The curriculum before 1975 was based on mathematical strands such as algebra, geometry and trigonometry (Zulkardi, 2002). This curriculum was mainly criticized for its inadequate attention to the relationship between different topics of mathematics (Ruseffendi, 1979). In 1975 a new curriculum was designed inspired by modern mathematics or 'new math'. However, it was criticised for using a structuralist approach to learning and lacking in real-life applications. The teaching resources of the new mathematics curriculum for primary schools originated from a project in Africa (Uganda) called *"Entebbe Mathematics Series"*. This was designed by university lecturers from the USA and England, while the teaching resources for junior and senior high schools came from a project in Scotland called *"Modern Mathematics for Schools"* designed by a collaboration of university lecturers and teachers in Scotland called *Scottish Mathematics Group (Depdikbud, 1977; Wirasto, 1972)*. In general, the approach used in *Entebbe Mathematics Series* was more formal than that used in *Modern Mathematics for Schools*.

In 1984, changes were made to the content structure of the curriculum. For example, statistics, which was previously only taught in senior secondary school, was moved to junior secondary school and introduced in primary school as well. Computers and calculators were introduced. In the revised curriculum of 1994, content in primary schools was extensively changed with a focus on teaching students how to count and use numbers in a set of algorithms. Relating to this, more emphasis was placed on number sense in primary school and graph theory was introduced in senior high school. Also, deductive thinking was incorporated in geometry. However, this curriculum was criticized for its heavy load. These criticisms resulted in another revision in 1999, which had a reduced number of mathematics topics. For instance, graph theory was removed from secondary mathematics.

In 2004, the government authorized schools to develop their own curriculum and tailor it to their local content. This school-based curriculum (KTSP) required staff and principals to be involved in the process and use their knowledge and skills to ensure the successful implementation of the curriculum (Zulkardi, 2002). Learning theories of constructivism and cognitive development were incorporated into the curriculum 2004 and problem solving ability, logical, critical, and creative thinking were emphasized. 2013 marked the introduction of yet another new curriculum, which is still in effect. This curriculum design was based on the diversification principle relevant to education units, local potential, and students with an aim to maintain the unified state of Indonesia (Alwasilah & Furqon, 2010). The principles followed by this curriculum include improved character; increased potential, intellect, and interest of students; diversity of local potential and environment; local and national development demands; employment demands; development in science, technology, and arts; etc.

Mathematics textbooks

Finkelstein (1951) reported shortages of textbooks during the independence consolidation and if they were available, few could afford to buy them.

In the Dutch colonial era, most textbooks were written by Dutch writers, including textbooks for Melayu language and local languages, with a consequence that the language of those books did not sound natural for Indonesian children (Nasution, 1983, p. 38). In the early independence era, many books written by Dutch writers were still used after they had been translated into Indonesian language. This was especially true for mathematics textbooks and other exact sciences books.

According to the regulation specified by the Government, however, school textbooks may be written by anyone who was interested and regarded as competent, but they must have been in accord with the curriculum and approval for them must have been obtained from the Government before they were printed and distributed to the public (Marsaban, 1968). For example, in order to implement Curriculum 1994, the Government (Department of National Education) gave opportunities to the public to write textbooks for primary and secondary schools and provided some guidelines for the writing. The draft of each book had to be submitted to the Government for evaluation, and if it was judged to meet the Government's criteria, then the book could be printed and distributed to the students for use in the classrooms. At that time, one of the writers of this article wrote a series of mathematics textbooks for junior high schools (Suwarsono, 1994).

However, when the Government considers that an incoming curriculum contains many new elements compared to the previous curriculum, or uses a new teaching approach with which most teachers are still not familiar, usually the Government (the department or ministry of education) decides to provide the textbooks for the first implementation of the curriculum, and all schools are obliged to use those books. This regulation has been enacted several times, for example in the implementation of Curriculum 1975 (where many new topics had been introduced in the curriculum) and Curriculum 2013 (where a new set of competencies had been introduced and a new teaching approach, namely a scientific approach, had been advocated). One of the problems with textbooks that often occurs is that because of the big population of students in Indonesia and the vast area of the country, the distribution of the books throughout Indonesia is often slow and the books are still unavailable when they are needed in the classrooms.

Alwasilah and Furqon (2010) highlighted some drastic changes of the textbook policy such as (1) there is no longer monopoly by Ministry of National Education (MONE) on textbook writing, provision, publishing, and distributing; (2) students are recommended to buy the books from the bookstores, while the teachers are not allowed to get involved in the textbook business; and Ministry of National Education (MONE) and Ministry of Religious Affairs (MORA) are encouraged to buy the copyrights of selected textbooks so that they can be mass-produced for students at relatively low cost.

Mathematics teaching approaches

Many changes have occurred in society during the 20th and 21st centuries. To reflect these changes, educational policy and school curricula in Indonesia have also undergone modifications. Implementations of new curricula brought with them new methods of teaching mathematics. These new approaches to teaching mathematics were influenced by curricula from other countries.

Cara Belajar Siswa Aktif (CBSA)

CBSA (Students Active Learning) was introduced to replace the former method of teaching, which was a lecture style of passive learning with no student engagement. Ruseffendi (2006) stated that when CBSA was launched by the Project for Teacher Professional Development (Proyek Pengembangan Guru/P3G) in 1979, teachers were asked to guide students to construct new knowledge, conduct investigations, and play puzzles so that interaction among the teacher and the students in multiple ways could occur. The principles of CBSA were formulated by Indonesian lecturers and teachers who visited countries in Asia, Europe, America, and Canada to observe what was considered to be good teaching methods for teaching mathematics.

Realistic Mathematics Education (RME)

The next major reform to mathematics education in Indonesia was from a group of Indonesian mathematicians and mathematics educators in the 1990s who researched mathematics education in different countries and chose to develop an Indonesian form of Realistic Mathematics Education (RME) that was developed by the Dutch. The reason for this change was because mathematics education in Indonesia, since the introduction of modern mathematics of the 1970's, had become the procedural learning of mathematics. (Sembiring, Hadi, & Dolk, 2008; Hadi, 2012). The Indonesian name for RME is Pendidikan Matematika Realistik Indonesia (PMRI) (Sembiring, Hoogland, & Dolk, 2010; Hadi, 2012) and the underlying principles were to make mathematics a meaningful activity (Freudenthal, 1986).

In 2001, the Directorate General of Higher Education at the Ministry of National Education (MoNE) Indonesia supported the implementation of RME in Indonesia in collaboration with Dutch Consultants from the Freudenthal Institute-Universiteit Utrecht and APS-Dutch National Centre for School Improvement (van den Hoven, 2010). PMRI activities include conducting a joint Master Program (IMPoME) between Universiteit Utrecht and Universitas Sriwijaya and Universitas Negeri Surabaya; publishing mathematics text books for grades 1 and 2; and undertaking a Mathematics Literacy Contest (KLM) (Hadi, 2012; Johar & Amin, 2010; Sembiring, Hadi, & Dolk, 2008; Zulkardi, 2013).

Contextual Teaching and Learning (CTL)

The project of CTL was initially undertaken at State University of Surabaya (UNESA) lead by prominent Indonesian scientist (Professor Mohamad Nur) and mathematics education leader (Professor Soedjadi) in collaboration with the University of Washington. CTL engages students in activities that help them connect their academic studies to the context of real-life situations (Bern and Ericson, 2001; Hudson, 2008; Johnson,

2000). The implementation of Contextual Teaching and Learning (CTL) in teaching mathematics commenced in 2000/2001 at junior secondary school level (Depdiknas, 2003). In 2006 CTL was introduced into the Indonesian curriculum. It encouraged teachers to commence each lesson with an introduction that connects the mathematics content to a real life situation (contextual problem); using a contextual problem at the beginning of lessons, guiding the learner to gradually master mathematical concepts (BSNP, 2006).

Lesson Study

The implementation of Lesson Study in Indonesia commenced in October 1998 with the project IMSTEP (*Indonesia Mathematics and Science Teacher Education Project*) in an attempt to strengthen the Teacher Education Program at LPTKs (Teacher Training Institutions). This project was a collaboration between the governments of Indonesia and JICA-Japan (*Japan International Cooperation Agency*). Initially only 3 LPTKs (Indonesian Education University (UPI), State University of Yogyakarta (UNY) and State University of Malang (UM)) were involved. The goal of the project was to improve the quality of Mathematics and Science Education teaching in Indonesia through an integrated program of pre-service and in-service development (Hendayana, Sukirman, & Karim, 2007; Karim, 2006). The principles of Lesson Study are based around the notion that the school is a learning community. They emphasise the importance of the iterative process of designing, conducting the lesson, observing and reflecting on the process after the lesson. Lesson Study developed the principle of collegiality and mutual learning to improve the quality of learning (Saito, Murase, Tsukui, & Yeo, 2014). By 2016 lesson study had been disseminated to over fifty LPTKs and schools in over 10 provinces in Indonesia.

ELPSA (Experiences, Language, Pictures, Symbols, Application)

The ELPSA framework provides a structure for identifying how mathematical concepts and understanding are acquired and developed. It represents five learning components, namely: Experience, Language,

Pictorial, Symbolic and Application. The framework for Indonesian schools was developed by Prof. Tom Lowrie and has been used to construct lessons and develop teacher professional programs in Indonesia since 2012 in cooperation with the World Bank (Lowrie & Patahuddin, 2015a, 2015b).

In 2014 Lowrie and Patahuddin received an Australian funded DFAT grant to introduce ELPSA to the West Nusa Tenggara region. The program is to promote mathematics engagement and learning opportunities for disadvantaged communities in West Nusa Tenggara (NTB), Indonesia. Together with Indonesian mathematics education experts, IKIP Mataram, leaders of educational jurisdictions and teachers from all districts in NTB, several activities have been conducted such as developing lesson and professional development modules, conducting professional development workshops for Leading Teachers at NTB and at the University of Canberra, and disseminating professional development lead by Leading Teachers with support from mathematics teacher educators.

Mathematics students' performances

Students' mathematics performance is usually assessed based on three yardsticks including National Exam, International tests, and Mathematical Olympiad. The National Examinations provide a strong mechanism that tends to impact on classroom instructional practices and is considered as a key benchmark by which education authorities and the public judge the quality of school performance. National exams that are administered to grades 6, 9 and 12, serve a couple of purposes such as indicating the quality of Indonesia's national education, whether or not students can proceed from one educational level to another, and which schools need assistance in achieving the quality of national education (Permendiknas, 2009). There have been four main national exams in Indonesia since 1965 (i.e., Ujian Negara [UN]; Ebtanas; Ujian Akhir Nasional [UAN]; Ujian Nasional [UN]) where a passing grade determined by the government indicated which students pass the exam. This grade gradually increased from a minimum score of 3.01 in 2005 to a minimum score of 5.25 in 2008 (Permendiknas, 2009). However, since 2015 national exams have ceased to determine whether students can graduate. Passing grades are determined by the schools based on school exams.

The impacts of national exams have been vast. Attempting to ensure students pass these exams has created enormous pressure for students, teachers, schools and even district governments, as more time must be allocated for exam preparation. This has resulted in larger workloads for both teachers and students. Extra lessons in and/or outside of the schools are made available for students to learn techniques about how to answer multiple-choice questions. In addition, national exams have also been criticised for exam fraud and cheating issues such as using mobile phones to send answers to other students, giving the answer key, either openly or discreetly, and changing the answer on the answer sheet (Arifin, 2012; Media Kompas, 2016). School principals and teachers have been arrested on that case. To prevent cheating, National Exam question variation had increased for middle and high school, from one to five in 2011, and from five to 20 in 2013. Even then, cheating still occurs (Media Kompas, 2016). Therefore, some argue that National exams do not provide an accurate measure about an Indonesian student's real competency. The latest attempt from the government to address cheating problems was to develop an integrity index and conduct computer-based national exams. The integrity index was intended to measure the honesty of students in each city or district (Jakarta Post, 2016). In addition, approximately 1 million students from 4,402 schools took a computer-based national exam (UNBK) compared to the previous year where only 594 schools used computers (Tamindael, 2016).

TIMSS and PISA as two international tests have been also used to assess Indonesian students' mathematics performance. TIMSS collects information about the quantity, quality, and content of instruction. The test items are matched against those in the curriculum or syllabus such as an understanding of fractions and decimals and the relationship between them. Statistics show that Indonesia has been an active participant of TIMSS since 1999 (only Grade 8). However, the mathematics performances of students have been much lower than the International average (500) or below most other participating countries.

Indonesian students' standing on TIMSS is presented below (http://litbang.kemdikbud.go.id); (http://timssandpirls.bc.edu)

- In 1999, Indonesian students ranked 34 out of 38 countries.
- In 2003, Indonesian students ranked 35 out of 46 countries.
- In 2007, Indonesian students ranked 36 out of 49 countries.
- In 2011, Indonesian students ranked 38 out of 42 countries.

Indonesia also has participated in PISA since 2000. PISA assesses students' ability to use their knowledge in real-life challenges rather than focusing on students' ability to master a specific school curriculum (OECD 2003). The PISA mathematics literacy test asks students to apply their mathematical knowledge to solve problems set in various real-world contexts. To solve the problems students must use a number of mathematical competencies as well as a broad range of mathematical content knowledge. Unlike TIMSS that is administered to 4th and 8th graders, PISA is taken by 15 year-old students who are typically in Grades 9 and 10. Many of TIMSS and PISA items are in the form of constructed-response and open-ended items whereas the questions in national exams are in the format of multiple choice and there are around 40 items which take 2 hours to complete. It is important to note that prior to the 60's, the national exam format was in an essay form (non-multiple choice).

PISA surveys in 2000, 2003, 2006, 2009, 2012, and 2015 (https://www.oecd.org) suggest that the standing of Indonesian students' mathematics performance is as follows:

- In PISA 2000 study, Indonesia ranked 39th of the 41 participant countries
- In PISA 2003 study, Indonesia ranked 38th of the 40 participant countries
- In PISA 2006 study, Indonesia ranked 50th of the 57 participant countries
- In PISA 2009 study, Indonesia ranked 61st of the 65 participant countries
- In PISA 2012 study, Indonesia ranked 64th of the 65 participant countries

Interestingly, despite the low mathematics performances of Indonesian students, the survey results in PISA 2012 show that Indonesia (among 64 or 65) has the highest percentage of students (96%) who reported being happy at schools. However, there has not been study to explore further how the students perceive the meaning of "happiness".

As the figures suggest, Indonesian students have performed poorly in TIMSS and PISA. However, they have had considerable achievement in International Mathematical Olympiads and won gold, silver and bronze medals in international contests. Some of their success in international competitions is listed below.

- In 2009, primary school students won 73 medals (gold: 13, silver: 20, and bronze: 40) in four competitions in mathematics and science.
- In 2011, after 24 years, Indonesian team in the IMO held in Amsterdam, the Netherlands, ranked 29th out of 101 countries (2 Silver medals and 4 Bronze medals).
- In 2014, 14 students (SD, SMP, dan SMA) won a Gold medal in International Mathematics Contest (IMC) in Singapore.
- In 2016, Junior high school students won a gold, 3 silvers, 4 bronzes in Thailand International Mathematics Competition (TIMC).
- In 2015, Indonesian University students won 2 gold medals out of 75 Universities in the world in International Mathematics Competition for University di Bulgaria (http://nasional.kompas.com).

Appearing at local and international mathematics competitions is valued by the Indonesian society and as such, students are given the opportunity to travel and compete in competitions at all education levels (from elementary school to University level). However, whereas some people use the success rate of students as an evidence for their ability to perform well in competitions if they have adequate support, others believe more attention should be paid to non-Olympiad students as Olympiad winners are less than 1% of student population (N= 44,637,259). It is also important to highlight that the low achievement of Indonesia may also be influenced significantly by the gap between provinces in terms of the provision of education facilities.

Mathematics teacher education and professional development

This chapter will elaborate on teacher professional development in particular (in-service) since the description of Indonesian pre-service teacher education can be found in another chapter book written by Abadi & Chairani (to appear in the forthcoming volume on Muslim States).

Indonesian education, one of the largest and most complex education systems in the world, has undergone reform efforts (Chang et al., 2013; Raihani, 2007) including implementing a range of policy reforms to improve the teacher education system (Abadi & Chairani, 2018) and the quality of teachers (Human Development Department East Asia and Pacific Region, 2010). Low mathematics performances of Indonesian students in national and international assessments both in TIMSS and PISA (Mullis, Martin, Gonzalez, & Chrostowski, 2004; Stacey, 2011) reinforced the urgency of improving teaching qualities. Subsequently, in December 2005, the Indonesian Government passed The Teacher and Lecture Law (No. 14/2005) that was considered as a comprehensive bill to raise the quality of Indonesian teachers and their teaching (Chang et al., 2013). It was assumed that a definitive measure of success for this initiative would be improved student academic performances.

The Teacher and Lecture Law was a derivative of the National Education System Law that indicated the minimum qualifications teachers should have (i.e., at least 4-year Bachelor qualifications relevant to the course being taught), the teacher certification process, the implication of the certificate in the teaching authority, and offered a financial incentive to those who completed the certification process. The law also defined (a) the competencies required of teachers in four areas (pedagogic, personal, social, and professional); (b) their incorporation into national teacher standards; (c) the role of various ministry units and agencies in supporting teachers to reach these competencies; (d) the teacher certification process and the qualifications required for such certification; and (e) the conditions under which teachers could receive special and professional allowances (Alwasilah & Furqon, 2010).

Through the certification process, teachers must demonstrate four competencies. Pedagogic competence relates to the teacher's understanding of the students and the mastery of techniques in planning and implementing effective instructions. Personality competence is about the maturity and personality of a teacher. Social competence is related to the competencies to communicate and socialize in social life. Professional competence is related to the thorough mastery of subject content to be taught (Firman & Tola, 2008).

The certification process occurred either through a portfolio assessment or a 90-hour training course. In order to receive the certification, teachers were required to prepare a portfolio and include a collection of documents such as their professional trainings, teaching experiences, sample of lesson plans, etc. Teachers' portfolios were assessed by a committee from a number of universities and those who passed would be certified. If the portfolios did not pass the assessment, teachers would be required to attend a 90-hour training course including a pre and a post test. The assessment at the end of the course was based on four components (i.e., written examination; teaching; participation; and peer teaching) that formed 75% of the final determination toward certification. The remaining 25% was the score obtained from the portfolio assessment, if applicable. The final score was an indication that the participating teacher had met the standards in the four competency areas as referred to in the Teacher Law and subsequent regulations.

Teacher certification was not only intended to improve teachers' competencies and professionalism but also their welfare through their professional incentive. A certified teacher would be entitled to an annual amount equivalent to his or her base salary (essentially a doubling of income). This was a substantial government investment, given that teacher salaries were the largest expense of the Ministry of Education and Culture (Chang et al., 2013). The incentive provided by the government for certified teachers attracted more prospective students into teacher education. According to a World Bank report, teaching became a more attractive profession and teachers colleges and universities experienced increased enrolment (Human Development Department East Asia and Pacific Region, 2010). For instance, the number of students enrolled in education programs in Indonesia increased from 200,000 in 2005 to more than 1,000,000 in 2010 (Chang et al., 2013, p. 99). In addition, approximately 500,000 teachers were enrolled in education programs at the Open University (Chang et al., 2013, p. 100). Hence, it appeared that the new law made the teaching profession a competitive option for talented graduates.

However, there is no evidence to suggest that certification improved teachers' teaching or that the certification process and the professional allowance benefited student learning (Chang et al., 2013). One comprehensive study using quantitative and qualitative methods was conducted to investigate the impacts of teacher certification on student learning outcomes (Chang et al., 2013). They found that the students of certified teachers did not have better learning outcomes than students of uncertified teachers. They also asserted that there was no difference between certified and uncertified teachers in their subject matter and pedagogy assessment scores.

It has been over a decade since The Teacher Law was released and the Indonesian Ministry of National Education has made constant efforts to develop structures and encourage the cooperation of stakeholders, including universities, provincial and district education offices, schools and teachers to implement the law. In 2005, approximately 80% of the almost three million Indonesian teachers did not meet the minimum requirements of being certified teachers. The Government targeted all these teachers should be certified by the end of 2015. However, in 2016, approximately 54% of Indonesian teachers are not certified yet (jendela.data.kemdikbud.go.id). Furthermore, the Indonesian government has been criticised for having placed little emphasis on critically evaluating the efficiency and the effectiveness of teacher development programs as well as their impact on student learning outcomes. More attention has been paid to achieving outputs such as increasing the number of teachers trained and providing resources to schools such as textbooks. As a result, limited evidence can be found about what works and what does not work to improve student learning.

Informal teacher professional development

The effect of policy reform on the teaching profession has been huge. The reform has given new hopes to teachers; however, it has also resulted in many challenges for them. The political situation in Indonesia characterized by the release of the new curriculum in 2013 and the changes of certification process have also caused much uncertainty for teachers on what policymakers are trying to achieve. This situation can be adequately captured through social media such as Facebook.

Facebook is extremely popular in Indonesia with over 64 million users who actively access their accounts monthly (The Jakarta Post, 2013). The ease of access to Facebook through mobile phones even in remote parts of Indonesia enables teachers to share experiences including their own challenges, questions, and frustrations, and seek support within the uncertain political situation marked by limited educational resources. In other words, much of the engagement on Facebook arises because of the challenges that Indonesian teachers face in getting access to resources and professional development. For example, many teachers have to individually source their own information and in doing so, turn to Facebook. Interestingly, some conversations on Facebook include mathematical and pedagogical ideas and the posts often prompt discussions among the Facebook users and are shared widely through other Facebook walls and groups.

The prevalent number of Facebook groups reflects the pervasive use of Facebook by Indonesian mathematics teachers and teacher educators. The first author identified over 100 Facebook groups created for Indonesian educators with the number of subscribers to many of these groups achieving over 50,000 members (access date 22 September 2016). For example, there are Facebook groups for particular levels of schooling (e.g., Matematika Gembira Guru SD is a closed Facebook group for primary mathematics teachers); regional groups (e.g., Guru matematika se Jatim is a closed Facebook group for mathematics teachers in East Java, MGMP matematika SMP Kab Sidrap is a Facebook group for Mathematics Working Groups in one district); mathematicians and teachers (e.g., Matematika UNM is a closed Facebook group originating at the Mathematics Department in one university), teacher educators and teachers (e.g., P4TK matematika is a closed Facebook group originating in the National teacher training institute, Paguyuban Guru dan Dosen is a closed Facebook group with 60,800 members), and teachers (e.g., Suara Guru Nusantara (Teachers voice) has 28,200 members).

Hence, it appears that Facebook has become a new mechanism by which teachers transform their preferences from word-of-mouth communication into informal written communication. Therefore, it is not surprising to conclude that Facebook groups and similar forms of social media have potential to be utilised as an effective supplement to traditional teacher professional development. However, the potential of Facebook for mathematics teachers' professional learning has remained largely unexplored in Indonesia. Very little is known about how mathematics teachers can use Facebook for their own professional learning and how Facebook could be used to support mathematics teachers' learning.

In light of the widespread use of Facebook in Indonesia, Patahuddin and Logan (2015) investigated the emergence of a community of practice (CoP) (Wenger, 1999) within the Facebook environment. They examined the mathematical and pedagogical interactions that occurred between Indonesian teachers and/or educators within the Facebook environment through analysing teacher responses to four Facebook posts concerning the division of a whole number by a fraction. The four posts were not designed to initiate a CoP; however, the researchers found that the manner in which teachers responded to the posts suggested there was potential for Facebook to be used to support mathematics teachers working and learning through collaboration. Their study demonstrated that engagement in Facebook environments opened opportunities for teachers to engage in conversational writing as well as professional learning dialogue. In addition, Facebook provided a mechanism for teachers to develop a personal voice that might not have emerged in more traditional professional learning settings. In another study, Patahuddin and Basri (2015) examined the engagement of Facebook users toward open and closed mathematical tasks and found that the open tasks stimulated rich mathematical conversation among Facebook users including teachers.

Therefore, we argue that Facebook has enabled teachers in Indonesia to express their ideas and share their challenges. Facebook seems to open an opportunity for policy makers to understand teachers and students' challenges such as to seek evidence of what works and what does not work to improve student-learning outcomes. However, there has been little research on the impacts of teachers' or the public's perceptions in shaping the educational policy in Indonesia.

Research in mathematics education

In general, the development of educational research in Indonesia has not significantly increased over the past two decades. In 2012 approximately 6.3 per cent of academics working in universities (N= 209,830) throughout Indonesia managed to publish in national journals and even fewer, less than 1%, in international journals (MOEC, 2012). The number of Indonesian publications in international journals is much less compared to that of neighbouring countries such as Malaysia, Singapore and Thailand (OECD, 2015). It appears Indonesian researchers lack motivation and a willingness to write for reputable refereed journals. Furthermore, the language barrier, as most international journals require writing to be in English, is also a hindrance. Therefore, most publications are in Bahasa Indonesia in local journals (Ristekdikti, 2016b).

Currently, the Indonesian government through the Ministry of Research, Technology, and Higher Education (Ristekdikti) has been promoting and supporting Indonesian academics (particularly Doctoral and Master degree) to conduct research and publish their findings in both international and national journals through providing grants for these purposes as an incentive (Ristekdikti, 2016a; 2016b). Furthermore, Ristekdikti released a regulation (Number 44 year 2015) about National Standards for Higher Education (SNPT). It asserts that graduates from Master's Programs should publish their research findings in national journals while graduates from doctoral programs should publish their research findings in international journals. According to Ristekdikti, in 2016 there are 199,888 lecturers in Indonesia, 67.8 % of whom have a master's degree, and 13.54 % have a doctor's degree. It is intended that in the next six years, at least 20 % of lecturers in Indonesia will have a doctor's degree. This intention is based on the desire that the number of international publications produced by Indonesian lecturers will increase by having more lecturers with a doctor's degree (http://forlap.dikti.go.id/dosen/homerekapjenjtetap).

The Head of LIPI (Indonesian Institute of Sciences) stated that efforts to encourage research and the dissemination of research findings should be via seminars and conferences organised in Indonesia and by creating Indonesian national and international journals (Zulkarnain, 2016). The Indonesian Mathematical Society (IndoMs) is making efforts to increase the availability for its members to publish their research findings by conducting annual seminars and conferences, in conjunction with universities, and by creating new mathematical journals that are peer reviewed. Also a number of Mathematics Education Departments from various universities conduct regular Seminars for mathematics education to increase the dissemination of mathematics education in Indonesia. IndoMs also publishes an international journal called IndoMs-JME (Journal on Mathematics Education). Currently, more than thirty national journals have been created. The considered top 3 are: Journal of Indonesian Mathematical Society on Mathematics Education (IndoMs-JME), Jurnal Ilmu Pendidikan (JIP) and Jurnal Pengajaran MIPA (JPMIPA). These are nationally accredited by Ristekdikti.

Concluding Comments

It can be concluded that during the pre-colonial era, Indonesia already had its own culture where education also existed as a process of passing local wisdom to the younger generations. This was conducted through the existence of *padepokans* during the Hinduism-Buddhism era, and *pesantrens* or *madrasahs* in the era of Islam. The main aim of this education was to teach young generations religious beliefs as well as basic necessities of life.

With the presence of colonial powers, especially the Portuguese and the Dutch, formal schooling was introduced. During Portuguese colonization, mathematics (arithmetic) had been introduced for the first time in schools. However, it is evident that the main aim of schooling was to provide Indonesian children with a certain level of knowledge and skills and who then could be recruited for the colonial offices or industry and to spread the colonial rulers' religious beliefs. Only a small proportion of Indonesian children (specially influenced by the Indonesian child's parents' wealth or position in the government) obtained an education. Another characteristic of education in Indonesia under the colonial powers was the non-existence of a coherent educational policy for Indonesian children. What already existed were just educational policies based on pragmatic needs or situational demands. This included the founding of Technische Hogeschool in Bandung, which was primarily established because of the difficulties Indonesian students faced in continuing their education after finishing their secondary education, as a result of the interrupted contact between Indonesia and the Netherlands resulting from the First World War.

It is also clear that mathematics (including arithmetic) was regarded as an important subject during the colonial period. This is evidenced by the fact that mathematics was taught in all school models, and the number of hours per week dedicated to the teaching of mathematics tended to be high, or at least higher than the corresponding number of hours taught for any other subject. We can also see from examples of mathematical problems used in examinations during that period that the mathematics that was studied was quite complex.

In the independence era (from 1945 onwards) the Indonesian Government has tried to set up a system of education on the basis of Indonesia's national conditions (Marsaban, 1968). However, it appears that intentionally or unintentionally the Indonesian system of education set up after independence has used the same system of education constructed by the Dutch colonial administration as its model, although it might use different names for the schools. For example, the senior high school (*Sekolah Menengah Atas, SMA*) was and is very similar to *Algemene Middelbare School (AMS)* set up by the Dutch colonial administration. The junior high school *(Sekolah Menengah Pertama, SMP)* was and is very similar to *Meer Uitgebreid Lager Onderwijs (MULO)*, and the primary school *(*earlier it was called *Sekolah Rakyat, SR,* and now it is called *Sekolah Dasar, SD)* was and is very similar to *Hollands Inlandse School (HIS)*.

With the large population size and lack of preparation in the establishment of an education system under the Indonesian Government, despite the use of the old Dutch system as a model, soon after the declaration of independence (August 17, 1945) Indonesia faced shortages of school buildings, shortages of teachers, and shortages of textbooks (Marsaban, 1968). The Indonesian Government has been steadily trying to fill the gaps. Nowadays, school buildings, teachers, and textbooks are generally no longer in short supply, except in some remote provinces, as teachers and textbooks sometimes are not evenly distributed. However, the quality of teachers and the quality of students' academic performances in general have caused great concern. Teachers' competence and students' performances still need to be improved which is indicated by the results of the *Uji Kompetensi Guru* (Teacher Competence Examination) and results shown by Indonesian students in international tests like TIMSS and PISA.

Along with the change of mathematics teaching materials, with the introduction of the New Mathematics Curriculum 1975, many new teaching approaches for mathematics have been introduced to Indonesia. Most of these approaches have used modern principles of teaching and learning mathematics such as student centered curriculum, contextual teaching and learning, realistic mathematics education, individualized learning, problem-based learning, and others. These new approaches have sometimes been promoted simultaneously with the introduction of new curricula (including Curriculum 2013, which contains some major changes to the approaches of teaching, learning, and evaluation). However, in terms of the Pedagogical practices of mathematics in the classrooms, not many changes seem to have occurred. One reason, perhaps, was the use of the national examinations at the end of primary school, junior high school, and senior high school. These tests were heavily concentrated on the coverage of many mathematical topics. This made teachers pay less attention to the implementation of new teaching and learning approaches and concentrate more on covering all the test topics in the time allocated to mathematics teaching.

Disssemination of new ideas on teaching and learning, and research in mathematics education has recently begun to increase in Indonesia, as indicated by the number of papers submitted for seminars and/or journals. However, the quality of the research still needs much improvement.

Despite the rather gloomy outlook above, one can also see some positive sides of the situation. The number of participants at seminars and conferences conducted to discuss new ideas in teaching and learning mathematics, the number of research papers submitted to seminars and/or journals, and the recently announced Government plan for the abolishment of national examinations, all indicate there is hope that in the future the situation can improve. However, hard work is still needed by all stakeholders for the betterment of mathematics education in Indonesia.

References

Abadi & Chairani, Z. (2018) Indonesia: The Development of Mathematics Teacher Education in Indonesia: Ch. IV in *Mathematics and its Teaching in the Muslim World, WSP Series on Mathematics Education*, Vol 13.

Alwasilah, A. C., & Furqon. (2010). Indonesia. In P. Peterson, E. Baker, & B. McGaw (Eds.), *International encyclopedia of education* (3rd ed.). Amsterdam: Elsevier Science.

Badan Pusat Statistik. (2016). Kepadatan Penduduk Beberapa Negara (penduduk per km2), 2000–2014. Retrieved from from https://www.bps.go.id/Subjek/view/id/12 - subjekViewTab3|accordion-daftar-subjek1

Bern, R. G & Erickson, P.M. (2001). Contextual teaching and learning. *The Highlight Zone Research @ Work* (5). 1–8

BSNP. (2006). *Standar isi mata pelajaran matematika*. Jakarta.

Arifin, H. (2012). Buku Hitam Ujian Nasional. *Yogyakarta: Resist Book dan CBE Publishing*.

Chang, M. C., Al-Samarrai, S., Ragatz, A. B., Shaeffer, S., De Ree, J., & Stevenson, R. (2013). *Teacher reform in Indonesia: The role of politics and evidence in policy making*: World Bank Publications.

Depdikbud. (1977). *Matematika untuk SMP Jilid 4 (Mathematics for Junior High Schools, Volume 4)*. Jakarta: Departemen Pendidikan dan Kebudayaan Republik Indonesia.

Depdiknas (2003) *Pendekatan kontekstual (Contextual Teaching and Learning)*. Jakarta: Departemen Pendidikan Nasional.

Finkelstein, L. S. (1951). Education in Indonesia. *Far Eastern Survey, 20*(15), 149–153. doi:10.2307/3023860

Firman, H., & Tola, B. (2008). The future of schooling in Indonesia. *Journal of International Cooperation in Education, 11*(1), 71–84.

Freudenthal, H. (1986). *Didactical phenomenology of mathematical structures* (Vol. 1): Springer Science & Business Media.

Hadi, S. (2012). Mathematics education reform movement in Indonesia *Proceeding of 12th International Congress on Mathematical Education* Seoul, Korea: ICME.

Hendayana, S., Sukirman, S., & Karim, M. A. (2007). Studi dan Peran IMSTEP dalam Penguatan Program Pendidikan Guru MIPA di Indonesia. [Study and Roles of IMSTEP in Strenghthening Science Teacher Education]. *Educationist, 1*(1), pp. 28–38.

Hudson, C.C & Whisler, V. R. (2008) Contextual teaching and learning for practitioners. *Systemics, Cybernnetics and Informatics, 6*(4). 54–59.

Human Development Department East Asia and Pacific Region. (2010). *Inside Indonesia's mathematics classrooms: A TIMSS video study of teaching practices and student achievement (Report No. 54936)*. Jakarta, Indonesia: The World Bank Office Jakarta.

The Jakarta Post. (2016). *Individual cheating more rampant in national exams. The Jakarta Post*. Retrieved 16 March 2017, from http://www.thejakartapost.com/news/2016/05/10/individual-cheating-more-rampant-national-exams.html

Jendela Pendidikan dan Kebudayaan. (2016). *Jendela.data.kemdikbud.go.id*. Retrieved 10 December 2016, from http://jendela.data.kemdikbud.go.id/jendela/

Johar, R., & Amin, S. M. (2010). Buku matematika PMRI kelas I SD sudah terbit. [Realistic Mathematics Book for Year 1 is published]. *Majalah PMRI, 3*(2), 25–26.

Johnson, E. B. (2000) *Contextual teaching and learning: What it is and Why It's Here to Stay*. USA: Corwin Press.

Karim, M. A. (2006). Implementation of lesson study for improving the quality of mathematics instruction in Malang. *Tsukuba journal of educational study in mathematics, 25*, 67–73.

Lowrie, T., & Patahuddin, S. M. (2015a). ELPSA–Kerangka kerja untuk merancang pembelajaran matematika [ELPSA - A framework for designing lessons]. *Jurnal Didaktik Matematika, 2*(1), 94–108.

Lowrie, T., & Patahuddin, S. M. (2015b). ELPSA as a lesson design framework. *Journal on Mathematics Education, 6*(2), 1–15.

Mailizar, M., Alafaleq, M., & Fan, L. (2014). A historical overview of mathematics curriculum reform and development in modern Indonesia. *Teaching Innovation, 27*(3), 58–68.

Marsaban, A. (1968). Indonesia. In T. W. G. Miller (Ed.), *Education in South-East Asia*. Sydney: Ian Novak Publishing.

Media Kompas. (2016). *Saling Contek Warnai Pelaksanaan UN di Gowa. KOMPAS.com*. Retrieved 16 March 2017, from http://regional.kompas.com/read/2016/04/04/13400071/Saling.Contek.Warnai.Pelaksanaan.UN.di.Gowa

MOEC (2012), *Performance Accountability Report of the Ministry of Education and Culture for 2012*, MOEC, Jakarta.

Mullis, I. V., Martin, M. O., Gonzalez, E. J., & Chrostowski, S. J. (2004). *TIMSS 2003 International Mathematics Report: Findings from IEA's Trends in International Mathematics and Science Study at the Fourth and Eighth Grades*: ERIC.

Nasution, S. (1983). *Sejarah Pendidikan Indonesia (History of Indonesian Education)*. Bandung: Jemmars.

OECD/Asian Development Bank (2015), *Education in Indonesia: Rising to the Challenge*, OECD Publishing, Paris. http://dx.doi.org/10.1787/9789264230750-en

Patahuddin, S. M., & Logan, T. (2015). Facebook as a learning space: An analysis from a community of practice perspective. In M. Marshman, V. Geiger, & A. Bennison (Eds.),

The 38th Annual Conference of the Mathematics Education Research Group of Australia (MERGA) (pp. 485–492). Adelaide: The Mathematics Education Research Group of Australasia.

Patahuddin, S. M. and Basri, H. (2015). Respon pengguna *facebook* terhadap tugas matematika. (Responses of Facebook users to mathematics tasks). *Jurnal Didaktik Matematika 2(2)*. 1–15

Pranoto, I. (2016). Mencari kurikulum Sriwijaya (Finding the Curriculum of Sriwijaya). *Harian Kompas.*

Raihani, R. (2007). Education reforms in Indonesia in the twenty-first century. *International Education Journal, 8*(1), 172–183.

Ristekdikti (2016a) Panduan Pelaksanaan Penelitian dan Pengabdian kepada Masyarakat Edisi X. [Guideline for Research and Community Service]. Jakarta: Ristekdikti.

Ristekdikti (2016b) Panduan Pengajuan Proposal Insentif Artikel pada Jurnal Internasional. [Guideline of Writing Proposal for International Journal Articles]. Jakarta: Ristekdikti

Ruseffendi, E. T. (1979). *Dasar-dasar Matematika Modern untuk Guru (The Basics of Modern Mathematics for Teachers)* (third ed.). Bandung: Tarsito.

Ruseffendi (2006) Pengantar kepada membantu guru mengembangkan kompetensinya dalam pengajaran matematika untuk meningkatkan CBSA. [Introduction to help teachers developing mathematics teaching competencies]. Bandung: Tarsito

Saito, E., Murase, M., Tsukui, A., & Yeo, J. (2014). *Lesson Study for Learning Community: A guide to sustainable school reform*: Routledge.

Sembiring, R. K., Hadi, S., & Dolk, M. (2008). Reforming mathematics learning in Indonesian classrooms through RME. *ZDM, 40*(6), 927–939. doi:10.1007/s11858-008-0125-9

Stacey, K. (2011). The PISA view of mathematical literacy in Indonesia. *Journal on Mathematics Education, 2*(2), 95–126.

Sunarno. (1951). *Ilmu Ukur Ruang (Solid Geometry)*. Jogjakarta: Prapancha.

Suwarsono, S. (1994). *Matematika untuk Sekolah Lanjutan Tingkat Pertama, Jilid 1, 2, 3) (Mathematics for Junior High Schools, Volumes 1,2,3)*. Jakarta: Widya Utama.

Tamindael, O. (2016). *Natiional examination no longer stressful. Antaranews.com.* Retrieved 16 March 2017, from http://www.antaranews.com/en/news/104010/natiional-examination-no-longer-stressful

van den Hoven, G.H. (2010). PMRI: a rolling reform strategy in process. In R. K Sembiring, R. K., K. Hoogland & M. Dolk (Eds.), *A decade of PMRI in Indonesia* (pp. 51–66). Bandung, Utrecht: APS International.

Wenger, E. (1999). *Communities of practice: Learning, meaning, and identity*: Cambridge university press.

Wirasto. (1972). *Matematika Modern Sekolah Dasar (Modern Mathematics for Primary Schools)*. Jakarta: Departemen Pendidikan dan Kebudayaan Republik Indonesia.

Yeom, M., Acedo, C., Utomo, E., & Yeom, M. (2002). The reform of secondary education in indonesia during the 1990s: Basic education expansion and quality improvement through curriculum decentralization. *Asia Pacific Education Review, 3*(1), 56–68. doi:10.1007/BF03024921

Zulkardi. (2002). *Developing a learning environment on realistic mathematics education for Indonesian student teachers*: University of Twente.

Zulkardi. (2013). Designing Joyful And Meaningful New School Mathematics Using Indonesian Realistic Mathematics Education. *Southeast Asian Mathematics Education Journal 2013, 3*(1), 17–26.

Zulkarnain, I. (2016, March 3). Jumlah Peneliti dan Publikasi Penelitian Masih Rendah. [Number of research and publications are low.] *Republika*. Retrieved from http://www.republika.co.id/berita/koran/kesra/16/03/03/o3gbg810-jumlah-peneliti-dan-publikasi-penelitian-masih-rendah

About the Authors

Sitti Maesuri Patahuddin earned her PhD from the University of Queensland Australia in 2009. Her bachelor and master degree were completed in Indonesia in the area of mathematics education. From 1999–2012, Sitti was a lecturer at the Department of Mathematics, State University of Surabaya, Indonesia. She had opportunities to join three research fellowships or postdoctoral programs in University of the Witwatersrand, South Africa; Charles Sturt University, Australia; and University of Canberra, Australia. Currently, Sitti holds a position as an Assistant Professor in the Faculty of Education, Science, Technology and Mathematics, University of Canberra, where she has been managing a 4-year Indonesian education project funded by DFAT Australia. Her research interests include the use of technology to enrich mathematics learning and teacher professional development, assessment of teacher's mathematics content knowledge for teaching, and the use of social media for teacher professional development.

Stephanus Suwarsono earned his PhD degree in mathematics education in 1982 from Monash University, Melbourne, Australia. Prior to obtaining his PhD, he had already earned the BA degree in mathematics and physics education in 1970 and the Drs degree in the same field, in 1974, both obtained at Sanata Dharma University, Yogyakarta, Indonesia (previously named Sanata Dharma Institute for Teacher Training and Education). He had also earned the BEd degree in 1976 in mathematics education, also at Monash University, Australia. Stephanus' research interests include the role of visual imagery and spatial ability in mathematical thinking, teacher education, history of mathematics and mathematics education, and the relationship between mathematics and mathematics education and culture. Currently he is a professor in mathematics education in the Department of Mathematics and Science Education, Sanata Dharma University, Indonesia.

Rahmah Johar is a lecturer at the University of Syiah Kuala, Aceh, Indonesia. She has held this position since 1998. Rahmah's undergraduate, graduate, and doctoral degrees are all in the area of Mathematics Education, completed respectively in IKIP Padang (1994), IKIP Surabaya (1997), and Universitas Negeri Surabaya (2006), Indonesia. Since 2006 she has been actively developing and socializing Realistic Mathematics Education. Furthermore, she has been a local expert in a 4-year Indonesian education project funded by DFAT Australia since 2015. Currently, Rahmah holds a position as a coordinator of the Master of Mathematics Education Program at University of Syiah Kuala. Her research interests focus on teacher professional development in teaching mathematics including helping teachers to develop students' reasoning, students' character, democratic classrooms, using video lesson, and implementation of technology. She is an editor-in-chief of the Jurnal Didaktik Matematika (www.jurnal.unsyiah.ac.id/DM) and a reviewer of some national journals in Indonesia.

Chapter VII

JAPAN: The History and Outlook of Mathematics Education in Japan

Naomichi Makinae
University of Tsukuba

Introduction

The school system and education laws

In this chapter, Japanese mathematics education is described using the following 3 points. The first is the current situation such as the present school system, national curriculum, textbooks and teaching. The second is a historical consideration that summarizes the development of a national curriculum policy for mathematics, and the third point provides perspectives for ongoing revision of a new national curriculum of mathematics education.

The Current School Mathematics Program

School system

The laws that prescribe the Japanese education system are Kyouiku-Kihon-Hou [Fundamental Law of Education] and Gakko-Kyouikuh-Hou [School Education Law]. The Fundamental Law of Education is the decree by which an educational idea in Japan is set. The School Education Law is the decree that defines the school as an educational institution.

The educational purpose for Japan is set as the Fundamental Law of Education as follows:

Education should aim for the full development of personality and strive to nurture the citizens to be sound in mind and body and imbued with the qualities necessary for those who form a peaceful and democratic state and society.

For this purpose, the following six principles are stated, which are respecting academic freedom, securing equal opportunity of education, guaranteeing lifelong education, guaranteeing compulsory education, and ensuring political and religious neutrality. Under these principles, the regulation and objectives in school are set by the School Education Law.

Compulsory education consists of 6 years of elementary school and 3 years of junior high school. After junior high, there are 3 years of high school and 4 years of university education that are not compulsory. Japanese school starts on April 1 and ends on March 31. Children from 6 to 11-year-old in April attend elementary school, and 12 to 14-year-old junior high school. Since the percentage of school attendance is 99.9%, this means all children of school age go to school, except in unavoidable circumstances.

National curriculum and textbooks

In the School Education Law, the aims of the school education are shown. One of the aims is related to mathematics. It says, "To understand a quantitative relation correctly that we use in daily life, and to cultivate the basic ability to develop it." These are the legal grounds which mathematics education is bound by in compulsory schools.

All subjects' detailed descriptions are mentioned in Gakko-Kyouikuh-Hou-Seko-Kisoku [The Enforcement regulations of School Education Law]. In elementary school, school subjects consist of Japanese Language, Social Studies, Arithmetic, Science, Physical Education, Craft, Music, Home economics and Life. In junior high school, they are Japanese Language, Social Studies, Mathematics, Science, Physical Education, Arts, Music, Technical arts, Home economics and Foreign language. Arithmetic in the elementary school and Mathematics in the junior high school are compulsory school subjects. These subjects' educational standards are described in Gakusyu-Shidou-Yoryo [The Course of Study] set by MEXT (Ministry of Education, Culture, Sports, Science and Technology). In the Japanese school system, the Course of Study plays the role of a national curriculum.

The current Course of Study for elementary and junior high school was revised and announced in 2008, and for high school in 2009. In elementary school, it has been in place since 2011; in junior high school since 2012; and in high school starting in the first year since 2013. So, it is from 2015 that the teaching of all grades in all schools has conformed to current standards.

The textbooks that are used in schools must be officially approved by MEXT. Private textbook companies edit textbooks according to standards indicated in the Course of Study. Textbooks are revised mainly according to the revision of the Course of Study and they also have minor revision every 3 or 4 years. These revisions of textbooks are written by specialists of mathematics education such as professors in universities and expert teachers. Companies must submit edited textbooks to the MEXT for approval. After approval, school districts select textbooks from those approved. The adopted textbooks are offered to all students free of charge every year, using the national budget.

Teacher training

Universities are responsible for teacher training in Japan. Teacher licenses are necessary in order to become a teacher. Regulations concerning a teacher's license are stipulated in Kyouiku-Shokuin-Menkyo-Hou [Educational Personnel Certification Law]. A teacher's license can be obtained by any person who has acquired a predetermined credit unit at a university, or who passed an education professional examination. Almost all teacher graduates acquire credits for a license, while a small number acquire a teacher's license by passing the national qualification examination.

Elementary school teachers are general classroom teachers, so their licenses include all subjects. On the other hand, junior high school and high school teachers have a license for each subject, because theirs is a subject teacher system. To acquire each license, 59 credits including educational training in schools are required.

Students take a teacher recruitment examination in their year of graduating from university. Teacher recruitment examinations for public schools are conducted for each board of education of prefectures and cabinet designated cities. Many boards of education conduct a two-stage selection, using written exams and interviews. The first year of recruitment for successful candidates is a provisional one. After one year, if there is no problem, provisional recruitment will be converted to regular employment. During the first year, all new teachers are trained as beginners. The content of the training course is given through lectures by teachers in leadership positions, while Lesson Study and discussion sessions among new teachers under training are conducted by the local Board of Education.

All teachers have to renew their licenses every 10 years. To renew a teacher's license, teachers need to take a course certified by MEXT. Lectures are held at a university, with a university faculty member who is in charge of teacher training serving as a lecturer. The license renewal course aims at acquiring knowledge about new educational administration, educational methods, subject content and maintaining teacher quality. Teachers pay for the course themselves and join the course during weekends, holidays or vacations.

Besides such official training for in-service teachers, teacher training is conducted for each school, district, and prefecture. Selecting the study theme each year, they conduct Lesson Study and listen to lectures by invited lecturers. Some enthusiastic teachers also participate in academic societies and research meetings.

Elementary program

According to the Course of Study for Arithmetic, there are 4 school hours per week for 1st graders, and 5 school hours per week for 2nd grade to 6th grade. 1 school hour is equivalent to 45 minutes. In the Course of Study, there are the statements about subject standards, consisting of subject objectives, and specific objectives and teaching content for each grade, including a brief syllabus design and suggested ways of teaching the content.

Objective

The objective of Arithmetic is stated as follows:

> Through mathematical activities, to help pupils acquire basic and fundamental knowledge and skills regarding numbers, quantities and geometrical figures, to foster their ability to think and express ideas with adequate insight and logic on matters of everyday life, to help pupils find pleasure in mathematical activities and appreciate the value of mathematical approaches, and to foster an attitude to willingly make use of mathematics in their daily lives as well as in their learning.

This objective may be divided into four teaching points. The first is to acquire basic and fundamental knowledge and skills. This point has been treated as a basis of math education from the past. And from an evaluation viewpoint, knowledge and skills are separated, knowledge is set with understanding and skill is set with operation and measurement. Second, to foster the ability to think and express with good perspectives and logically on matters of everyday life. This is a matter about children's thought. It's indicated as the mathematical thinking, judgment and the mathematical representation from an evaluation viewpoint. The third teaching point is for students to find pleasure in mathematical activities and appreciate the value of mathematical approaches, and to foster an attitude. While the previous points are based on a cognitive perspective, the third is about the affective side. Together, they should support understanding, rather than showing that child's affectivity is the important side where learning of mathematics is supported and emphasized. The fourth teaching point is about 'mathematical activities' and 'Through mathematical activities' is placed at the beginning of the objective. 'Mathematical activities' are concerned with the arithmetic on which children work independently with their own purpose. It means that learning of mathematics isn't passive learning by one-sided initiation from a teacher, but it should be an autonomous activity of children.

Teaching content

In order to indicate the teaching content systematically, four categories are set in the Course of Study and the content is divided into categories, that are: A. Numbers and Calculations, B. Quantities and Measurements, C. Geometrical Figures and D. Mathematical Relations. The teaching content for the elementary school is put into the order shown in Table 1 of the appendix to this chapter (p. 270).

This content should be taught at all schools, and it is expected that all students will be proficient in it in the same way as proficiency in Arithmetic is expected in elementary school. This means the contents of the Course of Study are minimum essentials for school mathematics. Teachers can teach matters outside what is written in the Course of Study according to the situation of children.

Two characteristics of the Arithmetic curriculum are continuity and spiral system. For example, in category A. Numbers and Calculations, as the grade progresses, the range of numbers increases and numbers are extended from integers to fractions and decimals, and students are supposed to master each arithmetical operation. In category B. Quantities and Measurements, after studying squares and rectangles in 2nd grade category C. Geometrical Figures, students determine areas of them in 4th grade. After studying parallelograms, rhombuses and trapezoids in 4th grade category C. Geometrical Figures, students determine areas of them in 5th grade. The connection among categories is also taken in consideration. In 4th grade, students study multiplication and division of decimal numbers in cases where multipliers and divisors are integers. Then in 5th grade they study multiplication and division of decimal numbers. This requires calculations of the same structure across two grade levels and is arranged with the intention of expanding the way of thinking about multiplication and division. These are examples of the spiral curriculum.

This content should not be taught as mathematical facts but as things that students understand through their own activities. For example, for a 5th grade to investigate features of geometric figures, such as parallelograms, rhombuses and trapezoids, tessellating them on a plane is mentioned as a mathematical activity.

Textbooks and teaching

Recent textbook pages are very colorful, and illustrations and photographs are often used. They intend to direct children's interests to learning Arithmetic. At the beginning of the chapter, concrete life events related to the content are presented. Then a new problem is presented. It is intended that students will construct a new concept, principle or law of mathematics through solving the problem. After that, exercises are presented. Figure 1 shows an actual page of the textbook used in school.

These are parts of a page of an introduction of addition with carry forward in 1st grade. At first, the situation where two children go to pick up nuts in autumn is presented. Next, a problem in the situation is presented as Yuka (girl's name in Japanese) picked up 9 nuts, and Hiroshi (boy's name) picked up 4. How many of them did they collect together?

After the problem, it is required to write expression '9 + 4 ='. Hiroshi says, '1...9,10,11,12,13. The total is 13.' He finds the answer 13 by counting. Yuka says, 'It is troublesome to count every time.' This is an introduction to new learning content, addition with carry forward, 9 + 4. Then 'Let's think about how to calculate' next. After turning the page, a way of thinking about this question is shown. Figure 2 shows the page.

There is explanation on how to find the answer of 9 + 4 with figures of blocks and a schema of calculation. As the first step, 9 needs 1 more to make 10. Second step, split 4 into 1 and 3. Third step, add 1 to 9 and make 10. Fourth step, 10 and 3 make 13. This way shows the idea of making 10 by compensating for the shortfall.

What is written on this part is for organizing and confirming what it is hoped that students thought about the question on the previous page. If they do not turn over the page, they cannot see the way of thinking asked on the previous page.

The concept of the textbook does not focus on teaching and practice calculation, but on students' thinking. Finding a way of calculation with students' mathematical activity is emphasized. By counting on, as had been studied before, students found the answer to 9 + 4 is 13 at first. Besides that, the problem becomes asking how to calculate 9 + 4. After summarizing the activity, the question 'Explain how to calculate 9 + 3' is presented. It is not asking for the answer to 9 + 3.

The word 'Textbook' in Japanese does not imply that teachers should use the contents of a textbook like a recipe, but rather teach using the contents of the textbook. Teachers are also expected to teach according to the actual situation of students. It is possible to change the problem situation, change the numerical value used in the problem, and change the order of contents. These can be the points of 'Jugyo-Kenkyu' [Lesson Study].

Many classes in Arithmetic are taught in a problem-solving style. When conducting Lesson Study, almost every teacher uses this style. In problem-solving classes, only one problem is solved in one class hour. The problem should conform to the aims of the lesson and be examined in detail after consideration and contemplation of its appropriateness and mathematical value. First of all, students solve the problem by themselves. After that, they compare and think about each way of thinking in the whole classroom, and they summarize what they studied at the end. This is typical of problem-solving type lessons. Textbooks are devised so that lessons can be taught using this approach.

Figure 1. Elementary textbook (Fujii T. & Iidaka S., Eds, 2011, p. 95)

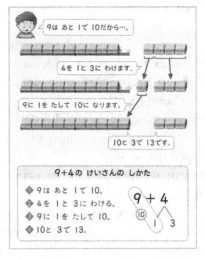

Figure 2. Elementary textbook (Fujii T & Iidaka S., Eds, 2011, p. 96)

Secondary program

For secondary education, Japan has a three-year junior high school and three years for (senior) high school. Mathematics is a subject in junior high, and there are six separate subjects under Mathematics in high school mathematics. In this chapter, the case of junior high school is mainly described as compulsory education. According to the Course of Study, for Mathematics in Junior high, 4 school hours per week for 1st (7th) and 3rd (9th) graders, and 3 school hours per week for 2nd (8th) graders are allocated. 1 school hour is actually equivalent to 50 minutes. In the Course of Study of junior high school, the statements are based on the same structure as for the elementary years. The statements for each subject consist of an overall objective, objectives and teaching content for each grade and brief advice on syllabus design and the way of teaching.

Objectives

The objectives of Mathematics for secondary students are stated as follows:

Junior High School

Through mathematical activities, (the aim is) to help students deepen their understanding of fundamental concepts, principles and laws regarding numbers, quantities, geometrical figures and so forth, to help students acquire the way of mathematical representation and processing, to develop their ability to think and represent phenomena mathematically, to help students enjoy their mathematical activities and appreciate the value of mathematics, and to foster their attitude towards making use of their acquired mathematical understanding and ability for thinking and judging.

High School

Through mathematical activities, (the aim is) to help students deepen the systematic understanding of basic concepts and mathematical principles and laws, to help students develop their ability to mathematically consider and express phenomena, cultivate the foundation of creativity, appreciate the value of mathematics, and to foster their attitude to make judgments based on mathematical argument by positively applying mathematics.

It is possible for this pair of objectives to be divided into five teaching points. The first is to understand fundamental concepts, principles and laws in mathematics. Especially in high school, it is required to understand these systematically. Students should not only understand separate mathematical facts but also begin to systematize them. The second aim is to acquire ways of mathematical representation and processing. Only in junior high is mathematical skill mentioned as an elementary objective. The third teaching point is to foster the ability to think and represent phenomena mathematically. Especially in senior high school, there is added need to cultivate the foundation of creativity. These are matters regarding how students think. It is indicated as the mathematical thinking, judgment and the mathematical representation from evaluation viewpoint. The fourth is to appreciate the value of mathematical approaches and to foster a positive attitude. While the previous three points stand on a cognitive perspective, the fourth is about the affective side. The point being made is that rather than confining mathematics learning to the classroom, students should try to use their mathematical thinking and mathematical logic as the basis for making decisions. The fifth is about 'mathematical activities'. The phrase 'Through mathematical activities' is placed at the beginning of the objectives. 'Mathematical activities' are concerned with the mathematics on which students work independently with their own purposes, as for Arithmetic in elementary.

Teaching content

In order to indicate the teaching content systematically, four categories are set out in the Course of Study for junior high and the content is divided into four categories: A. Numbers and Algebraic Expressions, B. Geometrical Figures, C. Functions and D. Making Use of Data. The teaching content for Junior high school is shown in Table 2 of the appendix to this chapter (p. 273).

This content should be taught at any school, and it is expected that all students will be proficient in the same Arithmetic as in elementary school.

The contents that the Course of Study indicates are minimum essentials for school mathematics. Teachers can teach matters other than what is written in the Couse of Study according to the situation of the students.

The characteristic of the Mathematics curriculum is to extend the range of numbers studied at first. In elementary school, students study only in positive number. When they enter junior high, they start to study negative numbers. In this time, they can use four fundamental operations of rational numbers too. After studying square roots, they can use all real numbers. Another characteristic is in the order of degrees of polynomials, equations and functions. In 7th and 8th grade, only linear equations and functions are treated; then in 9th grade, quadratic equations and special cases of quadratic functions are taught. Not only are higher levels of mathematics treated with the progress of each grade, but also the content of teaching is more systematized out of mathematical necessity. To solve a linear equation by formal processing, it is necessary to use rational numbers including negative numbers. Then, after extending the range from rational to real numbers, they can study quadratic equations and functions, and applications of the Pythagorean theorem. Finally, mathematical proof is introduced in 8th grade, dealing with Geometrical Figures. Proof, here, is not taught based on Euclidian Elements. Proof is taken up as a way to explain deductively, based on what is already known. Also, through the experience of proving, students are intended to explore the properties and relationships of the geometrical figures.

For high school mathematics, there are six subjects shown in Table 3 of the appendix to this chapter (p. 275). Two subjects required for graduation are Mathematics I or Applications of Mathematics. In almost every school, students take Mathematics I. Few students take Applications of Mathematics, and many schools do not offer it. Mathematics II, Mathematics III, Mathematics A and Mathematics B are optional subjects. Mathematics A is usually taken with Mathematics I in 10th grade. Mathematics II and Mathematics B are usually taken in 11th grade. Mathematics III is usually taken in 12th grade by students who aim to specialize in the mathematical sciences. The selection of subjects depends on what specialization students have chosen. Students who want to undertake literature or humanities at university need to take Mathematics I, A, II and B for the entrance examination.

Textbooks and teaching

In junior high school, the content to be studied is more abstract than in the elementary school. To overcome this challenge, textbooks are intended to direct students' interests to learning mathematics and mathematical activity. At the beginning of each chapter, concrete objects in daily life related to the content to be learned are presented typically by illustrations and photographs. Then a new problem is presented. It is intended that students will be assisted to construct a new concept, principle or law of mathematics through solution of the problem. There are usually several solutions, not just one.

Figure 3 shows an actual page of an 8th Grade textbook. In the page, a specific problem of mathematics is given for studying the properties of the sum of interior angles of polygons.

In the first page of the chapter 'Parallel and Congruent' in 8th grade geometrical figures, the theme 'Let's examine the properties of polygonal

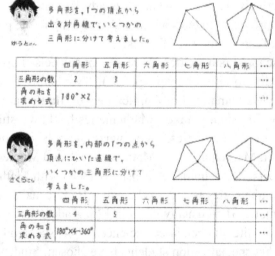

Figure 3. Junior high school textbook (Fujii T & Matano H., Eds, 2012, p. 90)

corners' is presented in larger letters than the title of the chapter. Under the theme, we can see several pictures that can be seen as polygons in real world, pieces of Shogi (Japanese Chess), and a goal net of soccer, some coins of the UK and the roof of a building. In this introduction, attention is directed to the types of polygons to be learned about in this chapter. Pentagons, hexagons, heptagons, and octagons can all be seen.

In this page, two students have appeared and explain each way of thinking about the sum of interior angles of polygons. The first is Yuto's (boy's name) idea. He divides the polygons into several triangles with diagonal lines coming out from one vertex. A quadrilateral is divided into two triangles, and the expression for calculating the sum of angles is $180° \times 2$. A pentagon is divided into three triangles. The column for calculating the sum of angles is left blank. In the cases of the hexagon and heptagon, both are also blank. Students are expected to complete these on their own.

Second is Sakura (girl's name), who divides a polygon into several triangles with straight lines drawn from one point inside the polygon to each vertex. The quadrangle is divided into four triangles, and the expression for calculating the sum of angles is $180° \times 2 = 360°$. The pentagon is divided into five triangles. The column of expression for calculating the sum of angles is blank. In the cases of hexagon and heptagon, both are also blank, and students are expected to write on them on their own. After Sakura's idea, Yuto asks why she subtracts 360° and Sakura says that it becomes more difficult to draw a figure as the number of polygonal corners increases.

On the next page, three instructions are given.

1. Using Yuto's idea, make an expression to calculate the sum of interior angles of an n-sided polygon with n vertices.
2. Using Sakura's idea, make an expression to calculate the sum of interior angles of an n-sided polygon with vertices.
3. Compare the expressions you made and write what you understand.

For such questions, it is intended for students to create their own formulas and to compare and consider diverse ideas. After these questions, the formula for the sum of interior angles of an n-sided polygon is given as $180° \times (n-2)$. Practice exercises are given after such content in the textbook.

The lessons based on such a textbook are expected to follow the problem-solving style the same as in elementary school, but this style is not always taken in all high school classes. There are also many classes where teachers explain the solutions written in the textbook, then let students practise questions and answer together. In high schools that focus on examination instruction guidance, this trend is particularly strong at present.

Development of Mathematics Education from an Historical View

Empiricism and Essentialism

Starting the New Education System of Post War Period

With the end of World War II in 1945, Japan was placed under the control of the Allied Forces. Japanese society was changed dramatically under the occupation in order to foster a democratic and peaceful nation. An educational system for the post war age was established at this time and the same system has continued until today. At that time, one of the main issues of school was changing to child-centered teaching in school. Pre-war and during the war, education had been considered to use uniform and unilateral teaching.

In 1947, when the Fundamental Law of Education and the School Education Law were enacted, a new educational system of the 6-3-3 year system began, dividing the years of elementary school, junior high school and senior high school. In the School Education Law, Mathematics as a school subject was established as Arithmetic in elementary school subjects, and as Mathematics in junior high school and high school. Standards for each subject were outlined in the Course of Study, and in the same year Shogakkou Tyugakkou Gakusyu Sidou Yoryou Sansuu-ka Sugaku-ka hen Sian [Tentative Course of Study for Arithmetic in Elementary School and Mathematics in Junior High School] were published by the Ministry of Education. The objectives of Arithmetic and Mathematics were described as the following:

> To develop students' ability to understand ideas of number, quantity and figures, to inquire and process various phenomena in daily life and to develop scientific attitude (Ministry of Education, 1947).

Following that, 20 concrete items of Arithmetic and Mathematics that the student should learn were presented. And "Sunsu [Arithmetic]" and "Tyuto Sugaku [Secondary Mathematics]" were published as new national textbooks. Despite being newly edited after the war, these textbooks were not consistent with the standards indicated in the Tentative Course of Study. This was an example of the post-war "chaos period".

In addition, in 1948, the "Sansu Sugaku-ka Shido Naiyo Ichiran-Hyo [Content List of Arithmetic and Mathematics]" was published and the standards that the list indicated were lowered by approximately one grade from the pre-war period. As new textbooks consistent with this standard, for the fourth grade of "Shogakusei no Sansu [Arithmetic for Elementary School Student]" and for the first grade of "Tyugakusei no Sugaku [Mathematics for Junior High School Student]" were published in 1949. Since these textbooks were published in the transition period from nationally produced textbooks to the national textbook screening system, they are regarded as textbooks approved by the Ministry of Education. Since then they are said to be models of approved textbooks published by private textbook companies. Here, teaching mathematics through solving problems in children's lived experience was introduced.

In response to these reforms, in 1951, Shogakkou Gakusyu Sidou Yoryou Sansuu-ka Hen Sian [Tentative Course of Study of Arithmetic in Elementary School] and Tyugakkou Koutogakkou Gakusyu Sidou Yoryou Sugaku-ka Hen Sian [Tentative Course of Study of Mathematics in Junior High School and High School] were published. The general purposes of Arithmetic and Mathematics shown here included inculcating attitudes to utilize mathematics in relation to daily life, attitudes to think and act voluntarily, and to show the usefulness and beauty of mathematics.

Empiricism in post war period

A *Unit Learning* method is based on close and daily observations of students' experience and interests, and drawing on these in lesson planning in all areas. Thus, teaching is contextualized in relation to children's experience or day-to-day activities. This method was widely adopted in the post-war period. Hence, a curriculum structure that incorporated daily life, not just math as a subject, was also introduced into mathematics education. The chart shown in Figure 4 is from a Tentative Course of Study of Mathematics (1951) for Junior High School and High School, and describes mathematics instructional content based on "living experience."

This curriculum model reflects the criticism against pre-war mathematics education, which overly emphasized teaching mathematics as a discipline and did not correspond with the children's real learning conditions. It was claimed that in order to incorporate children's needs in mathematics education, examples and problems discussed needed to be

Living Experience	Understanding and Skills	Vocabulary
3. Use proportion and inverse proportion to understand natural and social phenomena and solve problems Examples: Understand how to distribute main resources based on achievements and degree of importance, in order to understand national government policy and provide cooperation. Understand how interest rate is proportional to term and principal in the simple-interest method. When investigating factory management, to estimate cost price, consider that labor costs are proportional to wage bases.	Utilize and understand the features of proportional change in practical examples Express proportional relations in formulas and use those formulas to solve problems Understand and utilize the relationship between the constant of proportion and the slope of a linear proportional graph	Slope

Figure 4. Teaching Contents of Mathematics in Junior High School in 1951 (Junior high school grade 3 [Grade 9]). (Ministry of Education, 1951)

derived from everyday situations so that children could learn mathematics through solving those problems. The Allied Occupation Forces made this claim to the Japanese Ministry of Education, and this idea was also supported by a group of core curriculum proponents from the USA who had been promoting research in the field of pedagogy regarding the curriculum in their country. In addition, it was claimed that the standard of mathematics education did not match the children's comprehension ability, and after 1948, standards were lowered by one level in all grades, and instructional content that met the new standards was included.

The basis of this idea was to realize the daily usefulness of mathematics and to seek a way to teach practical mathematics rather than treating it solely as a preparatory step for studying advanced mathematics or merely as an exam subject to enter higher schools. It was thought more important for children to understand the necessity of mathematical skills based on one's experience rather than to hope they would someday understand the concepts. It was necessary to provide opportunities for children to learn the new content while understanding the necessity of mathematics based on their experiences. The instructional content shown in the chart above is structured in such a way that it lists living experiences and describes mathematical content related to those experiences. This represents the fundamental understanding of empirical mathematics education.

Using number expansion as an example, we could understand mathematics curriculum based on living experience in the following way. During the lower grades in elementary school years, children's lives are limited to the home and school environment, and the necessary numbers to understand their surroundings are limited to one to two-digit natural numbers. As the children enter into higher grades, they would be exposed to larger society such as community or country and would need larger numbers to understand their surroundings. In addition, measurements become sophisticated, and decimals and fractions become necessary. In lower secondary school, the children will need to understand the concept of square roots to capture phenomena expressed in quadratic equations and functions. Thus, the scope of numbers expands from natural numbers to rational numbers and to real numbers. As children develop, the scope and quality of their living experiences change, and the mathematics required also becomes more advanced. The empirical curriculum organizes its instructional content based on this principle.

Criticism of Essentialism

These initiatives to promote unit learning came under criticism after the conclusion of the San Francisco Peace Treaty, along with the objection to the occupation policy, and declined rapidly. The strongest point of the criticism was the decline in children's academic standards. Compared with surveys on computational problems conducted before the war, the results of the survey conducted at the time showed a clear drop in the academic standards among children. Of course, there certainly must have been many other factors responsible for the fall in academic standards such as the period of turmoil in education during and after the war, reduction in teaching content, and changes in the grade system. However, unit learning was targeted as the chief cause of the fall in children's academic standards.

Because unit learning takes up mathematics in relation to daily life, attention may not be focused on the mathematical content or may include perspectives or content other than that related to mathematics. Critics claimed these points became a major obstacle to teaching mathematics. In addition, if the curriculum is compiled based on children's lived experiences, mathematics may not necessarily be learned in a way that conforms to the traditional system. In fact, mathematics may be taken up in an order that does not fit the structure of mathematics, or some content that ought to be associated may not even be taken up. Based on these points, unit learning came under criticism, regarded as a haphazard teaching method that ignored the structure of mathematics, or that it just gave lessons on living instead of teaching mathematics.

In reaction to the criticism, teachers and schools asserted the significance of student-centered education and the necessity of solving problems. In 1958, however, on the occasion of the revision of the Course of Study for elementary school and junior high school, and the revision of high school in 1960, the instructional content was presented based on the system of subjects, and since then, systematized learning was adopted from essentialism (in the sense of pure mathematical principles).

However, this revision was not simply a change in policy against the Unit Learning method based on the empirical principle. The bigger aim was to revise the level of the teaching content. What the Ministry of Education and mathematics educators at the time saw as the problem in the Allied Occupation policy was the lowering of the standards of teaching content. What they were most interested in at the time was to raise the standard of the 1951 Tentative Course of Study that had reduced the pre-war teaching content back to the original standard as much as possible, while taking into account the actual situation of the children. Therefore, a systematic subject-based curriculum was adopted instead of a curriculum based on the empirical principle. If based on empiricism, priority is placed not on the assumption that the content of mathematics that ought to be learned will become useful some day in the future, but rather on the fact that they are necessary in a situation that students are currently facing. It is often mixed up with the notion of practicality in the familiar sense. Here, practicality refers to a condition in which something is operating and functioning on the spot, and it is asserted that students should learn mathematics that is actually useful in this sense. It does not mean simply mastering mathematics that students can utilize and apply. Furthermore, it does not mean that if students take a mathematical approach, they can discover relevant mathematics in their daily lives. Rather, it means to capture mathematics that exists in our daily lives, or on the contrary, to capture our daily lives that exist because of the existence of such a function. This is about learning mathematics that exists "on top of" or to explain people's experiences rather than mathematics that exists as an abstract idea in our minds. The objective here is totally different from the content as in the case of criticism against unit learning, and it is also not restricted to the principle of teaching that humans first gain recognition and understanding by experiencing. It is an issue of how to look at mathematics itself as a subject of learning.

There are characteristics to relating daily life with mathematics that can be viewed as actual textbook material. It has a slightly different meaning from the relationship between mathematics and daily life that was asserted in pre-war mathematics education. When taking up mathematics from events in daily life, the perception of life here is not restricted to its relationship with mathematics, but its relationships with other fields also come to be acknowledged. Life as a human activity is a broad concept, and its relationship with mathematics is part of a multifaceted view. At the same time, it takes on significance by being positioned within the whole. Mathematics education should not be viewed by restricting it to a study of mathematics as a perfected academic discipline; instead, mathematics should be considered as being related to various views and thoughts that exist in human activities. What was sought was learning this kind of mathematics.

Originally, empiricist mathematics education referred to recognizing mathematics as a part of these human activities and organizing it through children's experiences. It was not a matter of methodology regarding how to teach the instructional content. As intrinsic characteristics, empiricist mathematics education includes ways of approaching mathematics and focusing attention on the significance of mathematics that can be acquired from this point. These ways of approaching and thinking about mathematics are indeed what is desirable for the students to acquire.

New Math and its basis

The structure of mathematics in New Math

The advancement of science and technology in the 20th century was totally different from that up to the 19th century, and the level is much higher. After World War II, in a way, the world has developed through competition in development of science and technology. In contrast, school education is based on the conventional academic frameworks and teaching content, rarely reflecting academic development on subject content. It can be construed that the New Math movement was an attempt to reflect academic achievements of mathematics on such school mathematics that remained unchanged.

New content such as set theory, algebraic structure and linear transformation was incorporated in the mathematics of secondary education. The issues faced are not limited to how the new teaching content can be related to conventional ones and grade distribution. Other issues included structural changes and improvement of the mathematics curriculum as a whole specifically, the idea of set theory was to interpret the conventional learning on the range of numbers and the meaning of operations from a different perspective, which required changes in elementary school mathematics. The curriculum of New Math was incorporated into the revision of the Course of Study for elementary school in 1968, junior high school in 1969 and high school in 1970.

In order to justify such an improvement, the theoretic ground of curriculum was based on "The Process of Education" by J. S. Bruner (1960), in which he suggested the possibility to incorporate modern academic developments into school education with the hypothesis that "any subject can be taught effectively in some intellectually honest form to any child at any stage of development." (Bruner, 1960, p. 42) He suggested a spiral curriculum to elevate the level by analyzing the basic principle and structure of highly developed academic content, dealing with them in accordance with the developmental stage and learning repeatedly. It was believed that this would enable schools to teach mathematical structure at the earliest possible stage and to incorporate highly advanced content in school mathematics.

To summarize the concept of teaching content of mathematics, the significance of incorporating new teaching content is to learn the idea of New Math. This idea is the structure of mathematics. It was expected to enable students to understand the relations between different areas, grasp them from an integrated viewpoint and to make it easier to understand the elevated level of school mathematics.

"Back to Basics"

Teaching sophisticated mathematical concepts asserted in the New Math movement was a challenge not only for students but also for teachers. It was quite difficult to understand and teach the ideas of New Math. Moreover, since new advanced content was supposed to be handled in addition to maintaining traditional content, it often resulted in superficial instruction, which was far from the ideal. It was ridiculed as "Ochikoboshi Kyouiku [Dropout Education]" which neglected students' understanding, or "Shinkansen Jugyo [Super-Express Lesson]" which rushed ahead. So, critical examination of New Math was requested. Similar situations were seen in many countries that had advanced New Math. It was said that a "Back to Basics" shake back to the traditional teaching needed to occur.

Under these situations, the Course of Study for elementary school was revised in 1977, the junior high school in 1978 and the high school in 1979. Much content was reduced, especially that added in New Math. For such a selection, discussion occurred, with the theme of "What is basic content of Mathematics?". At that time, "Yutori Kyoiku [More relaxed education policy]" was introduced as a slogan for children to learn. It was against giving too much content to learn.

But it was not what mathematics educators wished for. Although it was true that the teaching of New Math was not a success, they still thought that the idea of New Math was important for mathematics education. To make students think logically, in an integrated way, and to learn mathematics developmentally, the structure of mathematics could be used as a good viewpoint for organizing mathematical facts, principles and methods. The ideas in set theory and algebraic structure could be seen as content. These same people argued that schoolteachers and general society did not understand the aims of New Math.

Academic achievement

"Zest for Life"

The next revision of the Course of Study was in 1989, and all curriculum levels were revised in this year. This revision was slightly different from the revision concerning traditional teaching content, as it was revised with an emphasis on the aims and methodological aspects of teaching.

The post-war rebuilding period and the high economic growth period were the age when society as a whole sought richness and economic development, and school education was also supported by the demands of these societies. On the other hand, when society as a whole becomes enriched, if it becomes possible to live a comfortable life without much effort or hardship, the reasons for what to learn will be thin. As a result, the motivation that supported learning at school was overly inclined towards the demands of examinations and fell into passive learning. In order to improve this situation, attention was paid to the subjectivity of the students themselves and their willingness to learn. It was claimed to nurture children's interests, motivation and attitudes, such as learning by themselves, getting to know the roles that mathematics plays in the world, appreciating the value of mathematics and making better use of what they have learned.

With the arrival of an advanced information society, changes in society will progress at a rate that is not comparable to conventional society. In such a rapidly changing society, it is important to update their own knowledge even after students graduate from high school, and to cultivate well-rounded persons who can think and act on their own. In Arithmetic and Mathematics, it is emphasized not only to acquire knowledge and skills, but also to enable thinking and judgment. And it reflects the view of new academic achievement including interests, motivation and attitudes such as self-study and self-thinking.

After that, the Course of Study for elementary school and the junior high school were revised in 1998. In 1999, the high school curriculum was revised. This revision corresponded to the 5-day school week, and a considerable reduction of the school hours, so a careful selection of the teaching content was done. (Before the revision in 1998, Japanese schools ran 6 days per week, with students attending Monday to Friday and Saturday mornings.)

In this same revision was the development of "Ikiru Chikara [Zest for Life]" as a balance of knowledge, virtue and body to live in a society undergoing change. It was introduced in the report of Tyuou-Kyoiku-Shingi-kai [Central Council for Education] in 1996. Zest for Life was explained as follows:

> It was clear to us that what our children will need in future, regardless of the way in which society changes, are the qualities and the ability to identify problem areas for themselves, to learn, think, make judgments and act independently and to be more adept at problem-solving. We also felt that they need to be imbued with a rich sense of humanity in the sense that while exercising self-control, they must be able to cooperate with others, have

consideration for their needs and have a spirit that feels emotion. It also goes without saying that if they are to lead vigorous lives, a healthy body is an indispensable requirement. We decided to use the term zest for living to describe the qualities and abilities needed to live in a period of turbulent change and felt it is important to encourage the right balance between the separate factors underlying this term. (Tyuou-Kyoiku-Shingi-kai, 1996)

In Arithmetic and Mathematics, "mathematical activity" was indicated in the objective of the subjects. Attention is focused on making students enjoy the fun and fulfillment of learning, and this point has also been passed down to the current Course of Study. In order to nurture the "Zest for Life", the idea of mathematical activity was proposed as a process of problem solving that enabled students to act on their own initiative.

International achievement tests

Although attention is paid to the emotional aspect of students, a focus is placed on their achievement. Based on the results of international surveys, the achievement of Japanese students in mathematics is highly regarded. But there were many criticisms about declining achievement around the revision in 1998 and 1999. Looking at the trends of results of TIMSS in Table 4 and PISA in Table 5, there was a period where the degree of achievement was lower.

The revisions of 2008 and 2009 were a response to such criticism. Before this revision, MEXT started Zenkoku Gakuryoku Gakusyu-joukyou Tyousa [National Assessment of Academic Ability] to grasp the nationwide achievement of students in 2007. This test has been held every year, except for the year of the earthquake disaster of 2011. The subjects and grades are Japanese and Arithmetic for 6th grade elementary school students and Japanese and Mathematics for 9th grade junior secondary school students. The test seeks to ascertain and analyze students' achievements and the learning patterns of school children from all of Japan and to investigate the outcomes of educational policies and programs, identify issues requiring attention, and make improvements therein. All public school students take this test.

Table 4. Results of mathematics in TIMSS (Kokuritsu Kyouiku Seisaku Kenkyu-jo, 2012)

Grade	1995	1999	2003	2007	2011
4th	567 (3/26)	Unexamined	565 (3/25) No significant difference	568 (4/36) No significant difference	585 (5/50) Significant rise
8th	581 (2/41)	579 (5/38) No significant difference	570 (5/46) Significant decrease	570 (5/49) No significant difference	570 (5/42) No significant difference

Table 5. Results of mathematical literacy in PISA (Kokuriku Kyouiku Seisaku Kenkyu-jo, 2013a)

2003	2006	2009	20012
534 (4/30)	523 (6/30)	529 (4/34)	536 (2/34)

In this test, an initial set of problems (Problems A) is given, described as "knowledge and skill problems". Then a second set (Problems B) is given, and these are described as problems requiring demonstration of an ability to judge what knowledge and skills are necessary to solve a practical problem and then proceeding to a solution. And the content related to the ability to put in practice, evaluate and improve the solution is included. Here, the quality of achievement required for school mathematics is indicated in the test. Figures 5 and 6 show several actual problems from the test.

New Curriculum for 2020

Learners' qualification and literacy

Behind the current Course of Study, revised in 2008 and 2009, there were criticisms of children's lowering achievement and amendment of the Fundamental Law of Education. Based on these, revisions have been made to use of the philosophy of "Zest for Life", and to return the hours of teaching and the teaching content to almost the same standard as that of 1989. However, this revision had not been done as a response to achievement issues but was undertaken while further discussing the purpose of traditional education. What attracted attention here was how to view the present times, with a knowledge-based society. Revisions were also guided by the viewpoint of OECD-PISA (Organisation for Economic Co-operation and Development - Programme for International Student Assessment).

[5] Mai and Koharu decided to investigate what kind of rectangles their classmates consider beautiful. They distributed a survey sheet with a 5 cm segment as shown below to their 33 classmates and asked them to draw a rectangle.

Figure 1 summarizes the results to the survey. From this histogram, we can say, for example, there are 5 rectangles with the horizontal length (width) greater than or equal to 2 cm but less than 3 cm.

Fig. 1: Distribution of Rectangles (Horizontal Length)

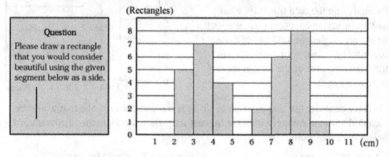

Answer the following questions (1) to (3)

(1) The rectangle Mai drew has the width of 8.2 cm. In fig. 1, it belongs to the category of rectangles greater than or equal to 8 cm but less than 9 cm. The drawing Koharu drew had the width of 3.1 cm. To which category in Fig. 1 does Koharu's rectangle belong?

The rectangle Mai drew and the rectangle Koharu drew are shown below.

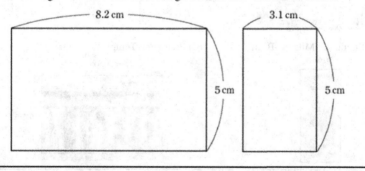

Figure 5. National Assessment of Academic Ability Mathematics B (Kokuritsu Kyouiku Seisaku Kenkyu-jo, 2013b)

(2) Mai thought if Kohara's rectangle were rotated, it would look similar to the one she drew.

Therefore, she determined the ratio of the longer side of the rectangle to the shorter side on each of the rectangles they collected, and then summarized the results in the histogram shown in Fig. 2.

By analyzing the data this way, what new information becomes clearer about which rectangles the students consider beautiful? Explain your answer using the characteristics of the histogram shown in Fig. 2.

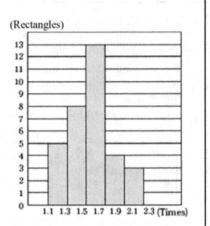

(3) If we consider the shapes of (a) to (d) as rectangles, there is one that would be included with the category with the highest frequency shown in the histogram in Fig.2. Select the correct one.

(a) Arc de Triomphe

(b) Book of "Taketori-Monogatari"

(c) Stamp of "Mikaeri-Bijin"

(d) Parthenon Temple

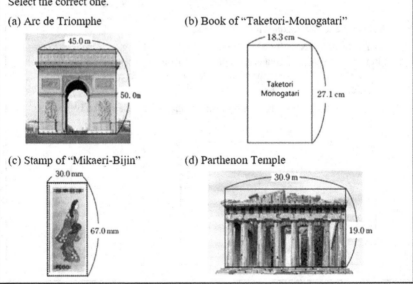

Figure 6. National Assessment of Academic Ability Mathematics B (Kokuritsu Kyouiku Seisaku Kenkyu-jo, 2013b)

In the 21st century, new knowledge, information and technology will have tremendous importance as the foundation of activities in all areas, including politics, economics, culture and the like, and expertise will be needed for children living in a knowledge-based society. This was the starting point for considering the purpose of education. It was also an international trend, and the OECD defined the necessary competence for children living in knowledge-based societies as key competencies. At the heart of the key competencies is thinking and acting thoughtfully and they are categorized in the following three categories. They are "Ability to utilize social, cultural and technical tools interactively", "Ability to form human relationships in diverse social groups" and "Ability to act autonomously". Based on this idea, PISA evaluates the ability to utilize learned knowledge and skills for real-life tasks, not mere fixed knowledge and routine skills, through distinctive questions that are different from conventional achievement test questions.

In PISA, mathematical literacy is the ability to be questioned about mathematics. Mathematical literacy is an individual's capacity to formulate, employ and interpret mathematics in a variety of contexts. It includes reasoning mathematically and using mathematical concepts, procedures, facts and tools to describe, explain and predict phenomena. It assists individuals to recognize the role that mathematics plays in the world, to make well-founded judgments and decisions needed by constructive, engaged and reflective citizens. (OECD, 2015, p. 65)

This cannot be said to be a completely new perspective for Japan, and the same goal can be read in the Mathematics in the Tentative Course of Study for junior high school and high school in 1951. In addition, in the "Zest for Life" that was the basis of the revision of the Course of Study in 1998 and 1999, the idea common to PISA key competencies was evidenced and the objectives of Japanese Arithmetic and Mathematics have been keeping them on the same track.

As a fundamental position of OECD's key competencies and mathematical literacy, there is a viewpoint on what kind of qualifications a child is to acquire. In the discussion of the purpose of education, it was emphasized that education should be based on learners' qualifications and literacies. While in the discussion of the purpose of mathematics education in the past, the focus was on what should be taught as mathematical content. In 2013, the National Education Policy Research Institute summarized in its report "Syakai no Henka Ni Taiosuru Shishitsu ya Nouryoku WO Ikuseisuru Kyoikukatei Hensei no Kihon Genri [Fundamental Principle of Curriculum Development to Foster Qualifications and Literacies to Respond to Changes in Society]".

The proposed "Competencies for the 21st Century" are shown Figure 7. Based on what is defined in foreign countries as a key competency and the framework of OECD-PISA etc., the qualifications and

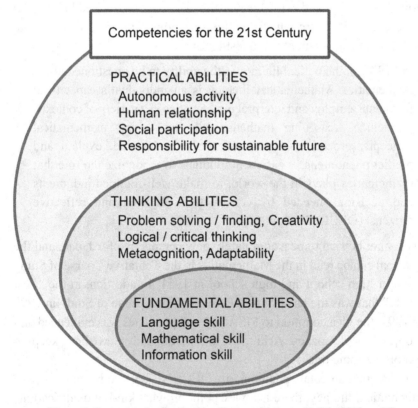

Figure 7. Competencies for the 21st Century (Katsuno Y. Ed., 2013, p. 26)

literacies to be developed are categorized as "basic literacies", "cognitive skills", and "social skills". These are organized into a three-layer structure (shown in Figure 7) with thinking ability as the core, fundamental ability to support it and practical ability to orient the usage of it.

In 2014, on the basis of the framework of qualifications and literacies, Kyoiku-Katei Kikaku Tokubetsu Bukai [Curriculum Study Group of Central Council for Education], an advisory board for MEXT, compiled the qualifications and literacies as three pillars for the foundation of the next curriculum.

The three pillars suggested to MEXT were:

1. What you know and what you can do (Individual knowledge and skills)
2. How to use what you know and what you can do (Thinking ability, judgment ability, expression ability, etc.)
3. How to live a better life and relate to society and the world (Individuality, diversity, cooperativeness, attitude to learn and humanity)

(Kyoiku-Katei Kikaku Tokubetsu Bukai, 2014)

Since 2016, the Council has been conducting revision work in each school subject for the next curriculum. Courses of Study will be revised soon, and will be applied in elementary schools in 2020, in junior high schools in 2021, and in high schools from 2022.

Mathematical perspective and thinking

In the process of revising, the working group of Arithmetic and Mathematics was set up and was requested to clarify the three pillars of qualifications and literacies for Arithmetic and Mathematics. For clarification, they set a viewpoint Sugaku-tekina Mikata Kangaekata [Mathematical Perspective and Thinking], which is an important characteristic of the subject.

Regarding Mathematical Perspective and Thinking, we can see the words "Sugaku-tekina Kangaekata [Mathematical Thinking]" in the objective for Arithmetic in the Course of Study for elementary schools revised in 1958 and 1968, Mathematics in junior high school revised in

1958 and 1969, and high school revised in 1960 and 1970. "Mathematical Thinking" has been positioned not only in the objective, but also as the viewpoint of evaluation. They say, in studying Arithmetic and Mathematics students can acquire knowledge and skills, and explore by utilizing the acquired knowledge and skills by using Mathematical Perspective and Thinking. Also, with regard to 'How to live a better life and relate to society and the world', students need Mathematical Perspective and Thinking. It works on all of the three pillars of qualifications and literacies, and plays a role as the core when students think logically, base judgments on sound grounds, or express their reasoning. As for Mathematical Perspective, it can be summarized as paying attention to the concept and also to the quantities, figures and their relationships and grasping their characteristics and essence. On the other hand, Mathematical Thinking can be summarized as thinking in an integrated or developmental way while looking back and associating knowledge and skills of the exercise when utilizing numbers, formulas, figures, tables, graphs, etc., according to purpose, thinking logically, and solving the problem. Based on these, Mathematical Perspective and Thinking cultivated in Arithmetic and Mathematics is described as thinking logically in an integrated and developmental way. Students need to focus on quantities, figures and their relationships.

Therefore, in the working group, the qualifications and literacies for Arithmetic and Mathematics were organized for each school stage as follows.

Arithmetic in elementary school

Through mathematical activities such as utilizing learning arithmetic for life and further learning by using Mathematical Perspective and Thinking, teachers aim to cultivate students' qualifications and literacies as follows:

1. Understand fundamental concepts, principles, rules of quantities and figures and acquire the skills to express and process everyday events mathematically.
2. Develop the ability to think logically about phenomena by using mathematics, to consider in an integrated way and developmentally

by finding properties such as quantity, figures, and to express phenomena concisely, clearly and accurately using mathematical expressions.

3. Appreciate the value of mathematics and acquire attitudes to utilize mathematics for life and further learning, look back on learning and solving problems better.

Mathematics in junior high school

Through mathematical activities by using Mathematical Perspective and Thinking, teachers aim to cultivate students' qualifications and literacies as follows.

1. Understand fundamental concepts, principles, rules of quantities and figures and acquire the skills to mathematize phenomena, interpret, express and process mathematically.

2. Develop the ability to think logically about phenomena by using mathematics, to consider integrally and developmentally by finding properties such as quantity, figures, and to express phenomena concisely, clearly and accurately using mathematical expressions.

3. Appreciate the value of mathematics and acquire attitudes to think persistently by utilizing mathematics, to apply it to life and further learning, and to look back on the process of problem solving, evaluate and improve it.

Mathematics in high school

Through mathematical activities such as thinking and clarifying the essence in mathematics by using Mathematical Perspective and Thinking, teachers aim to cultivate students' qualifications and literacies as follows:

1. Understand basic concepts, mathematical principles, and laws in mathematics systematically and acquire the skills to mathematize phenomena, interpret, express and process mathematically.

2. Develop the ability to think logically about events by using mathematics, to clarify the essence by looking back on the process of thought, to consider in an integrated way and developmentally,

and to express phenomena concisely, clearly and accurately using mathematical expression.

3. Appreciate the value of mathematics and acquire attitudes to think persistently by utilizing mathematics, to develop judgment based on mathematical rationale, and to review and improve the process of problem solving. (Sansu Sugaku Working Group, 2016)

The three qualifications and literacies described at each school level correspond to the three pillars such as "Knowledge and Skill", "Thinking, Judgment and Expression" and "Ability to learn expressiveness". "Knowledge and Skill" requires conceptual understanding, understanding of methods for problem-solving, skills to express and process mathematically, and so on. It is important to understand that even though 'knowledge and skills' is what is written, that means effective knowledge and skills to solve problems, and working effectively when thinking or judging. In "Thinking, Judgment and Expression", it is required to find problems and solve problems by utilizing knowledge and skills. For an "Ability to learn expressiveness", it is required to appreciate the value of mathematics and to think persistently and flexibly.

Process of mathematical learning

In order to cultivate students' qualifications and literacies, the role played by the learning process is extremely important. In Arithmetic and Mathematics, mathematical activity has been regarded as student learning processes. It has been emphasized that these processes are at the core of learning mathematics, in which students understand phenomena, find mathematical problems, solve the problem autonomously and collaboratively, including the process of forming a concept and systematizing the solution process, and look back on mathematics. Such a problem-solving process is important, and this process is mathematical activity.

As shown in Figure 8, this mathematical problem-solving process is based on mathematical understanding of everyday life and social events mathematically, expressing and processing mathematically, solving problems, and looking back on the solution process.

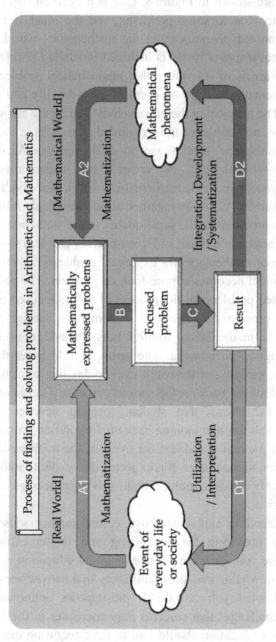

Figure 8. The mathematical problem-solving process (Sansu Sugaku Working Group, 2016)

Two cycles are shown in Figure 8. One is a cycle of mathematizing real-world phenomena to solve problems, and the other is a cycle of solving mathematized phenomena in the mathematics world. Problem solving in the mathematics world is included in solving problems in the real world. The arrow of A1 shows that mathematical problems can be created by mathematizing from an event of everyday life and society in the real world. The problem in the process indicated by the arrow of A1 is mathematical, so this problem can be regarded as matters in the world of mathematics. On the other hand, the arrow on the opposite A2 shows that mathematical problems can be created by mathematizing from mathematical phenomena in the mathematical world. Since both problems of arrow A1 and arrow A2 can be regarded as mathematical problems, they are expressed as problems of the same quality in the mathematical world.

B means to formulate a vision and a prospect for problem solving using mathematics about mathematically expressed problems. In this process, the original problem becomes a more focused problem. The process of C solves the focused problem. In the process of C, the focused problem is mathematically processed and solved in a manner appropriate to the purpose. From the mathematical solution thus obtained, consideration is given to the original phenomena. The arrow of D1 shows the student making clear the meaning of the result in everyday life, or utilizing it for other events in looking back on the solution process. On the other hand, the arrow of D2 shows that student forming new concepts and systematizing result through looking back on the solution process.

The cycle of the real world and the cycle of the mathematical world develop in relation to each other. It is expected that students autonomously and collaboratively conduct such problem-solving processes, and act on their own initiative.

Under the directions explained above, current mathematics education in Japan is in preparing for organizing and implementing a new curriculum. Here, the purpose of mathematics education is also not limited to the knowledge and skill familiarity in a narrow sense of what you know and what you can do. It presupposes unknown problem situations and challenges that students may encounter in the future, and how mathematics education should aim at their acquiring qualifications and literacies to solve new and unseen problems. A new practice of mathematics education commensurate with this is about to begin.

References

Laws in Japan

Kyouiku-Kihon-Hou [Fundamental Law of Education], 1947 Mar. 31 Law No. 25, Last amendment: 2006 Dec.12 Law No.120. (in Japanese).

Gakko-Kyouikuh-Hou [School Education Law], 1947 Mar. 31 Law No.26, Last amendment: 2016 May 20 Law No. 47. (in Japanese).

Kyouiku-Shokuin-Menkyo-Hou [Educational Personnel Certification Law], 1949 May. 31 Law No.147, Last amendment: 2016 May.20 Law No. 47. (in Japanese).

Gakko-Kyouikuh-Hou-Seko-Kisoku [Enforcement Regulations of School Education Law] 1947 May 23 Ministerial Ordinance of Ministry of Education No.11, Last amendment: 2016 Mar.31 Ministerial Ordinance of Ministry of Education, Culture, Sports, Science and Technology No.19. (in Japanese).

Official Documents

Kokuritsu Kyouiku Seisaku Kenkyu-jo [National Education Policy Research Institute] (2012). Kokusai Sugaku Rika Kyouiku Doukou-Tyousa 2011nen Tyousa Kokusai Tyousa Kekka Houkoku [International Results in Trends in International Mathematics and Science Study 2011], (in Japanese).

Kokuritsu Kyouiku Seisaku Kenkyu-jo [National Education Policy Research Institute] (2013a). OECD Seito no Gakusyu Toutatsu do Tyousa 2012 nen Tyousa Kokusai Kekka [International Results in Program for International Student Assessment 2012], (in Japanese).

Kokuritsu Kyouiku Seisaku Kenkyu-jo [National Education Policy Research Institute] (2013b). Zenkoku Gakuryoku Gakusyu-joukyou Tyousa Sugaku B [National Assessment of Academic Ability Mathematics B] (in Japanese).

Monbu-kagaku-syo [Ministry of Education, Culture, Sports, Science and Technology] (1998a). Shogakkou Gakusyu Sidou Yoryou [Course of Study in Elementary School], (in Japanese).

Monbu-kagaku-syo [Ministry of Education, Culture, Sports, Science and Technology] (1998b). Tyugakkou Gakusyu Sidou Yoryou [Course of Study in Junior High School], (in Japanese).

Monbu-kagaku-syo [Ministry of Education, Culture, Sports, Science and Technology] (1999). Koutogakkou Gakusyu Sidou Yoryou [Course of Study in Junior High School], (in Japanese).

Monbu-kagaku-syo [Ministry of Education, Culture, Sports, Science and Technology] (2008a). Shougakko Gakusyu-Shido-Yoryou [Course of Study in Elementary School] (in Japanese).

Monbu-kagaku-syo [Ministry of Education, Culture, Sports, Science and Technology] (2008b). Tyugakko Gakusyu-Shido-Yoryou [Course of Study in Junior High School] (in Japanese).

Monbu-kagaku-syo [Ministry of Education, Culture, Sports, Science and Technology] (2009). Koutogakko Gakusyu-Shido-Yoryou [Course of Study in High School] (in Japanese).

Monbu-sho [Ministry of Education] (1947). Shogakkou Tyugakkou Gakusyu Sidou Yoryou Sunsuu-ka Sugaku-ka hen Sian [Tentative Course of Study for Arithmetic in Elementary School and Mathematics in Junior High School], (in Japanese).

Monbu-syo [Ministry of Education] (1948). Sansu Sugaku-ka Shido Naiyo Ichiran-Hyo [Content List of Arithmetic and Mathematics], (in Japanese).

Monbu-syo [Ministry of Education] (1951a). Shogakkou Gakusyu Sidou Yoryou Sansuu-ka Hen Sian [Tentative Course of Study of Arithmetic in Elementary School], (in Japanese).

Monbu-syo [Ministry of Education] (1951b). Tyugakkou Koutogakkou Gakusyu Sidou.

Yoryou Sugaku-ka Hen Sian [Tentative Course of Study of Mathematics in Junior High School and High School], (in Japanese).

Monbu-sho [Ministry of Education] (1958a). Shogakkou Gakusyu Sidou Yoryou [Course of Study in Elementary School], (in Japanese).

Monbu-sho [Ministry of Education] (1958b). Tyugakkou Gakusyu Sidou Yoryou [Course of Study in Junior High School], (in Japanese).

Monbu-sho [Ministry of Education] (1960). Koutogakkou Gakusyu Sidou Yoryou [Course of Study in Junior High School], (in Japanese).

Monbu-sho [Ministry of Education] (1968). Shogakkou Gakusyu Sidou Yoryou [Course of Study in Elementary School], (in Japanese).

Monbu-sho [Ministry of Education] (1969). Tyugakkou Gakusyu Sidou Yoryou [Course of Study in Junior High School], (in Japanese).

Monbu-sho [Ministry of Education] (1970). Koutogakkou Gakusyu Sidou Yoryou [Course of Study in Junior High School], (in Japanese).

Monbu-sho [Ministry of Education] (1977). Shogakkou Gakusyu Sidou Yoryou [Course of Study in Elementary School], (in Japanese).

Monbu-sho [Ministry of Education] (1978). Tyugakkou Gakusyu Sidou Yoryou [Course of Study in Junior High School], (in Japanese).

Monbu-sho [Ministry of Education] (1979). Koutogakkou Gakusyu Sidou Yoryou [Course of Study in Junior High School], (in Japanese).

Monbu-sho [Ministry of Education] (1989a). Shogakkou Gakusyu Sidou Yoryou [Course of Study in Elementary School], (in Japanese).

Monbu-sho [Ministry of Education] (1989b). Tyugakkou Gakusyu Sidou Yoryou [Course of Study in Junior High School], (in Japanese).

Monbu-sho [Ministry of Education] (1989c). Koutogakkou Gakusyu Sidou Yoryou [Course of Study in Junior High School], (in Japanese).

Sansu Sugaku Working Group [Working group of Arithmetic and Mathematics Curriculum Study Group of Central Council for Education] (2016). Shingi no Torimatome [Summary of deliberations], (in Japanese).

Tyuou-Kyoiku-Shingi-kai [Central Council for Education] (1996), 21seiki wo Tenboushita Wagakuni no Kyouiku no Arikata ni tsuite [The Model for Japanese Education in the Perspective of the Twenty-first Century], (in Japanese).

Books

Bruner J. S. (1960). The Process of Education, Kenbridge: Harvard University Press.

Fujii T & Iidaka S. Eds. (2011). Atarasii Sunsu 1 [New Arithmetic 1], Tokyo: Tokyo Shoseki Co. Ltd. (in Japanese).

Fujii T & Matano H. Eds. (2012). Atarasii Sugaku 2 [New Mathematics 2], Tokyo: Tokyo Shoseki Co. Ltd. (in Japanese).

Katsuno Y. Ed. (2013). Syakai no Henka ni Taiosuru Shishitsu ya Nouryoku wo Ikuseisuru Kyoikukatei Henseino Kihon Genri [Fundamental Principle of Curriculum Development to Foster Qualifications and Literacies to Respond to Changes in Society], Tokyo: Kokuritus Kyoiku Seisaku Kenkyu-jo, (in Japanese).

Organisation for Economic Co-operation and Development (2016). PISA 2015 Assessment and Analytical Framework Science, Reading, Mathematic and Financial Literacy, Paris: Turpin Distribution Services.

About the Author

Naomichi Makinae is an associate professor at the University of Tsukuba, Japan. He is also concurrently serving as a commissioned member of the Japanese Ministry of Education, Culture, Sports, Science and Technology and a commissioned researcher of the National Education Policy Research Institute. As an academic activity, he is a vice chair of the editorial board of the Journal of the Japan Society of Mathematical Education. His research field is about the history of mathematics education. In his recent research, there are papers about mathematical education based on empiricism in postwar education reform and the origins of lesson study in Japan.

JAPAN: Appendix

Table 1. Teaching Content of Arithmetic in Elementary School

Grade	A. Numbers and Calculations
1	Meaning and representations of two-digit numbers and three-digit numbers in simple cases Addition of two one-digit numbers, and subtraction, and of two-digit numbers in simple cases
2	Meaning and representations of four-digit numbers Size and order of numbers by the decimal positional numeration system Addition of two-digit numbers, and subtraction, and three-digit numbers in simple cases Multiplication up to 9 times 9 and a two-digit number and a one-digit number in simple cases Simple fractions such as 1/2 and 1/4
3	Unit of ten-thousands ('man' in Japanese) Addition and subtraction of 3- and 4-digit numbers Multiplication of 2- or 3-digit numbers and 1- or 2-digit numbers Division where both the divisors and the quotients are 1-digit numbers Meaning and representations of decimal numbers to the tenths place Meaning and representations of fractions Addition and subtraction of fractions in simple cases Representations of numbers and simple addition and subtraction using soroban (Japanese abacus)
4	Units of hundred million ('oku' in Japanese) Round numbers Division in the cases where the divisor is a 1-digit or 2-digit number and the dividend is a 2- digit or 3-digit number Addition and subtraction of decimal numbers Multiplication and division of decimal numbers in cases where multipliers and divisors are integers Addition and subtraction of fractions with the same denominators

Grade	A. Numbers and Calculations
5	Even numbers and odd numbers
	Divisors and multiples
	Multiplication and division of decimal numbers
	Addition and subtraction of fractions with different denominators
	Multiplication and division of fractions when multipliers and divisors are integers
6	Multiplication and division when multipliers and divisors are fractions

Grade	B. Quantities and Measurements
1	Comparing length, area and volume
	Reading clock times
2	Meaning of units and measurements of length [m, cm, mm]
	Meaning of units and measurements of volume [L, dL, mL]
	Time [day, hour, minute]
3	Meaning of units and measurements of weight [g, kg]
	Units of length and time [km, second]
4	Meaning of units and measurements of area and determination [cm^2, m^2, km^2, a, ha]
	Meaning of units and measurements of angle [° (degree)]
5	Determinations of the area of triangles, parallelograms, rhombuses, and trapezoids
	Meaning of units and measurements of volume, and determinations [cm^3, m^3]
	Average of measured values
	Representation and comparing of quantities that are obtained as a ratio of two quantities of different types
6	Estimations of the area of shapes
	Determinations of the area of circles
	Determinations of the volume of prisms and cylinders
	Determinations of speed
	System of the metric units

Grade	C. Geometrical Figures
1	Recognizing the shapes of objects
2	Triangles and quadrilaterals Squares, rectangles, and right triangles
3	Isosceles triangles and equilateral triangles and angles Circles and spheres and center, radius and diameter
4	Parallelism and perpendicularity of straight lines and planes Parallelograms, rhombuses and trapezoids Cubes and rectangular parallelepipeds
5	Polygons and regular polygons Congruence of geometrical figures Ratio of the circumference of a circle to its diameter Prisms and cylinders
6	Reduced figures and enlarged figures Symmetric figures

Grade	D. Mathematical Relations
1	Representation of situations where addition and subtraction are used, by using algebraic expressions
2	Relationships between addition and subtraction Representation of situations where multiplications are used, by using algebraic expressions Organizing and classification of numbers and quantities in everyday life and representation using simple tables and graphs
3	Representation of situations where divisions are used, by using algebraic expressions Representation of numbers and quantities by using □ Organizing and classification of data, and representation by using tables and bar graphs
4	Algebraic expressions that contain some of the four basic operations and parentheses ( ) Representation of numbers and quantities by using □ and △ Properties of commutative, associative, and distributive laws Organizing and classification of data and representation by using broken-line graphs

Grade	D. Mathematical Relations
5	Proportional relationships in simple cases
	Percentage
	Organizing and classification of data and representation by using pie graphs and band graphs
6	Ratio
	Proportional relationships
	Inversely proportional relationships
	Algebraic expressions by using letters such as a and x
	Determination of the average of data and the distribution of data
	Analyzing all the possible outcomes systematically for actual events

Table 2. Teaching Content for Mathematics in Junior High Schools

Grade	A. Numbers and Algebraic Expressions
1 (7th)	Positive and negative numbers and the fundamental operations
	Algebraic expressions using letters
	Simple linear equations
2 (8th)	Addition and subtraction with simple polynomials, and multiplication and division with monomials
	Simultaneous linear equations with two unknowns
3 (9th)	Square roots
	Expand and factor simple polynomial expressions $$(a+b)^2 = a^2 + 2ab + b^2$$ $$(a-b)^2 = a^2 - 2ab + b^2$$ $$(a+b)(a-b) = a^2 - b^2$$ $$(x+a)(x+b) = x^2 + (a+b)x + ab$$
	Quadratic equations
	Solution formulae for quadratic equations

Grade	B. Geometrical Figures
1 (7th)	Parallel translation, symmetric transformation and rotational transformation Positional relationship between straight lines and planes in space Fundamental methods for constructing figures like bisector of an angle, perpendicular bisector of a line segment and perpendicular lines Representation of space figures on a plane Determinations of the surface area and volume of basic cylinders, pyramids and spheres
2 (8th)	Properties of parallel lines and angles Properties of angles of polygons Congruence of plane figures and the conditions for congruence of triangles Basic properties of triangles and parallelograms Necessity, meaning and methods of proof
3 (9th)	Similarity of plane figures and the conditions for similar triangles Properties of geometrical figures Properties of ratio of line segments related to parallel lines Relationships between the scale factor, the ratio of areas and the ratio of volumes of similar geometric figures Relationships between inscribed angles and its central angle in a circle Pythagorean Theorem

Grade	C. Functions
1 (7th)	Meaning of functional relationships Meaning of coordinates Representation of proportion and inverse proportion into tables, algebraic expressions and graphs
2 (8th)	Representation of linear functions into tables, algebraic expressions and graphs
3 (9th)	Representation of the function $y = ax^2$ into tables, algebraic expressions and graphs

Grade	D. Making Use of Data
1 (7th)	Arranging data into tables and histogram, and reading trends focusing on its representative values [mean, median, mode, relative frequency, range, class]
2 (8th)	Probability of an uncertain event in simple cases
3 (9th)	Sample surveys in simple cases

Table 3. Teaching Content of High School Mathematics

Subject	Content
Mathematics I	Expand and factor polynomial expressions Sets and logic Plane geometry Quadratic functions Data analysis
Mathematics II	Higher degree equations Geometrical figures and equations Exponential function and logarithmic function Trigonometric functions Calculus
Mathematics III	Quadratic curves Complex numbers Limits Differential calculus Integral calculus
Mathematics A	Probability Properties of integers Properties of geometrical figures
Mathematics B	Probability distributions Sequences Vectors
Applications of Mathematics	Mathematics and human activity Mathematical considerations in social life

Chapter VIII

SRI LANKA: School Mathematics Program of Sri Lanka, its History, Contemporary Status, and Future Plans and Aspirations

Lucian Makalanda
Teachers College Columbia University and Queensborough Community College CUNY

Sri Lanka: Some Background

Sri Lanka (see Figure 1) was known as Ceylon until 1972. Ancient Sri Lanka was a country that was based on agriculture, and the leaders of ancient Sri Lanka, its Kings, used many different applications of Mathematics to help rule the country. There were many changes for school Mathematics progress in Sri Lanka during and after the colonial period (1815–1948), and there were major political reforms during years 1972 and 1997.

Figure 2 shows a map of the Democratic Socialist Republic of Sri Lanka. Its administrative capital is Sri Jayawardenepura Kotte ("Kotte"), and its largest city is Colombo.

Figure 1. Map excerpt, showing Sri Lanka near the tip of India

277

Figure 2. Map of Sri Lanka　　　Figure 3. Typical Sri Lankan primary students

The number of government schools and students enrolled in them has increased more than threefold since 1950. Numbers that were around 3,200 increased to around 10,012 in 2013 according to the Data Management branch of the Ministry of Education Sri Lanka. The student to teacher ratio has decreased from about 35:1 to 18:1 and the adult literacy rate has increased from 65% to about 93%. The number of teachers has almost increased more than fivefold since 1950.

Portuguese (1505–1656) and Dutch (1632–1802) were a presence in Ceylon (the name of Sri Lanka before 1972), and the British ruled the country during 1796–1948. In 1833, under the Colebrooke-Cameron commission, local English schools were established, and the missionary schools that had previously taught in the vernacular also adopted English.

Dr. C.W.W. Kannangara, who was born in October 13, 1884 and died on September 23, 1969, was the "father of free education" in Sri Lanka. He served as the Minister of Education who introduced reforms to the education systems of Sri Lanka and who introduced the education

Figure 4. Dr. C.W.W. Kannangara

bill containing the free education scheme on May 30, 1944. This was passed into law as the British Education Act.

Since then, the changes in Mathematics education in Sri Lanka have had an enormous impact on the Sri Lankan population. It is notable that this first major change, introducing universal free education from kindergarten to university, was implemented three years before Sri Lanka obtained political freedom from its British colonial powers. Before the introduction of free education, Mathematics and Science had been taught almost exclusively in English medium schools, which served about 10% of the total student population of Sri Lanka at that time. English education was the pathway to higher education and better employment back then, hence the demand for education in general and English in particular was rising. Mathematics and Science also were becoming popular. There was a shortage of persons qualified in Mathematics and Science and the available few were concentrated in the urban areas.

The Government of Sri Lanka opened up schools in rural areas and recruited Mathematics and Science teachers by giving them higher salaries in order to meet these demands and rising pressures. Still, Mathematics and Science were taught almost exclusively in the English Schools in Sri Lanka around this time.

Education after Independence (1948)

After political independence in 1948, the second huge change was initiated in 1952 with the introduction of Swabasha (mother-tongue) schools. This resulted in the spread of learning in Mathematics and Science to about 20% of the student population. The demand for Science and Mathematics, however, continued to rise. The need for their teachers became greater with the existence of two types of schools—Government and assisted—the latter increasing in number as they were set up by individuals and private institutions. The appointment of full time Science teachers to many schools was unjustifiable due to the small number of students in these classes. The private assisted schools were the most reasonable choice for them because of their size and other luxuries that private schools offered during this period.

In 1961, the government took over the assisted schools while laying the foundation for a unified system of schools. The teachers were put to better use while eliminating the wastage. The number of students who were doing Science and Mathematics increased to more than 30%.

The force against the takeover was the Catholic Christians. The religious flavor in the schools would be lost, they argued, if the schools were taken over by the government. However, the Buddhist organizations and the leftist movements forced the takeover by arguing that the Christians, who were less than 10% of the total population, were enjoying more benefits than the majority Buddhist group, which formed about 65% of the total population.

Dramatic changes were expected with the takeover of the schools by the government. In 1957, two universities had been opened, however they produced more arts and oriental language graduates than Science and Mathematics graduates. Hence a change in the school curriculum was needed to produce more graduates in Science and Mathematics.

In Peradeniya, Sri Lanka, a British Commonwealth education conference was held in 1963. Afterwards, changes to the school curriculum in Mathematics and Science were developed by two committees appointed by the Director of Education. Mathematics was made compulsory for all the students in secondary schools, as recommended by the Mathematics committee. However, there was still a shortage of qualified Mathematics teachers and a need to reeducate a large number of existing Mathematics teachers. All teachers were also exposed to more pleasant ways of teaching, using new methodologies.

The introduction of these new methodologies required teachers and teaching aids that the government could not afford. Hence the teachers had to improvise and create new ways to assist their teaching. The principles of Mathematics education, the methodology of teaching Mathematics, basic knowledge of Mathematics, the psychological bases of Mathematics education and the preparation and use of teaching aids using locally available materials were some of the steps for which guides were designed for the teachers. The changes in the methodology were not reflected in the current textbooks, which were not satisfactory, with many now irrelevant and having too strong a focus on drill exercises. The Mathematics

Curriculum Committee had to produce a series of textbooks that were suitable for the new methodologies.

Famous mathematical books around 1965 in Sri Lanka

Mr. S. L. Loney (1860–1939), a professor of Mathematics at Royal Holloway College, Egham, Surrey, University of London, and fellow of Sidney Sussex College, Cambridge, was the author of many Mathematics textbooks that were used in Sri Lanka for many years after 1965. One of the books he authored was "The Elements of Statics and Dynamics", published in 1897.

Prof. J. E. Jayasuriya (1918–1990) wrote the first set of Mathematics textbooks used for many years by students in Sri Lanka after 1965. They covered topics such as Geometry, Algebra, Number systems and Number theory.

In 1965, new textbooks, in the form of Mathematics readers, which included exercises designed to help children to do individual study through discovery learning, began to be produced. They were first introduced into Grade 6. As a result of these changes, the proportion of Grade 6 students learning Mathematics rose to over 90% of the student population. The new textbooks were planned to be introduced to Grades 7 and 8 in subsequent years. However, after the 1965 general elections, the new program was introduced to Grade 7 only in 1967 and to Grade 8 in 1969. Sets, relations, symmetry, probability and statistics were introduced in a very elementary way in this program. The Science/Mathematics curriculum unit that did all this work was moved to a permanent curriculum development center in 1968.

The need for improvements in the educational system around the 1970's

After this stage, over 60% of the secondary school student population (Grades 6–11) were learning Mathematics. More and more people started realizing the importance of Mathematics and there was a general awakening that more Mathematics teachers were needed. Even though there was a substantial push during this period, there were many issues yet to overcome. Only about 50% of students had boxes of instruments, and even pencils and/or exercise books were lacking. Poverty, a variation in the availability of resources and, in some instances, carelessness, were some of the reasons that many students were not adequately prepared for the new curriculum.

Teachers' guides were also too bulky and rather unattractive in appearance. Some teachers were not sufficiently well prepared to begin using them in class and avoided doing so. All teachers were expected to read and study the guides but some had problems understanding them, which hindered their ability to use them in teaching their students. A consequence of this was that their students were less motivated, as the proper methods were not being used. Not surprisingly, some teachers found textbooks more convenient to use than the guides, but it was the guides that contained advice on different possible teaching sequences suggested by their designers.

To ensure that some students could be prepared for the UK General Certification of Education (GCE) and Ordinary level Examination (OLs), a new alternative syllabus in Mathematics was developed in 1970 and introduced into Grade 8 then and into Grade 9 in 1971. About 20% of it was devoted to new content, covering topics in logic, sets, relations, probability, statistics, matrices, number bases and number systems. This new syllabus was focused towards Science students, but alternative syllabi were considered for Commerce and Arts students and other majors. The first students took the GCE and OLs examinations in 1972.

After discussion with Heads of schools, only 47 out of an initially chosen 72 schools agreed to implement the new syllabus. Curriculum Development Center officers discussed problems that the teachers were having by meeting with them in each region and visiting every school once a term. With the help of the British Council, a consultant from the United Kingdom was obtained to assist in closely tracking the progress of the new syllabus in these schools. Also, draft notes were prepared and mailed to schools regularly.

It was necessary to change the primary school syllabus to better prepare students for this new secondary school syllabus. With the guidance of Prof. Geoffrey Mathews of Chelsea College, London, and Miss Joan

Bliss of Rousseau Institute, Geneva (later at Chelsea College), a child study program was commenced in 1968. This pilot study was carefully done in three stages. In 1975, the first group reached secondary school, followed by an extension to more schools, and finally in 1983 to reach all primary school students—about 8,000. Both rural and urban area schools were carefully selected for the pilot stage. Teachers involved in this were selected from consenting training college lecturers, in order to maintain the high standard of this program.

The 1971 Insurrection

These education reforms were seen as revolutionary, but high school graduates were accused by some of "knowing what to do, but not knowing how to do it", and even the very few who "knew how to do it" were accused of "not recognizing any need for it". Lack of communication among various participants exacerbated this discussion. Some argued that, while at that time there were large numbers of people with high levels of learning (and, presumably, with high aspirations), there was a national shortage of the craftsmen and technicians needed for the development of the country.

The number of unemployable educated graduates increased, but no one with government responsibility for addressing this problem took any action. The whole curriculum had to be drastically changed in order to cater for the need of the country to fix the increasing number of educated unemployed graduates. This resulted in young people coming into conflict with the Sri Lankan government and gradually organizing against it. In 1970, the existing government had to face a huge defeat in the general elections, and the newly elected government had a socialist outlook. The expected significant government changes in economic, social and education policy were not introduced immediately but were delayed for a whole year. During this period, the nation's youth were getting more organized with firearms and other types of weapons, and a rebellion against the existing Sri Lankan government in April 1971 resulted in many youths being killed or arrested. These serious outcomes led the government to develop both short and long term proposed solutions to these problems, to avoid the reoccurrence of such events in the future.

Redesign of the curriculum after the 1971 revolt

The whole curriculum was redesigned to fulfill dual purposes. It was decided that all students would sit for a public examination called the National Certificate of General Education (NCGE), after finishing nine years of general education. After doing so, 80% would leave school and join the work force of the country, while the remaining 20% would pursue higher secondary education, leading to the Higher National Certificate of Education (HNCE).

At that time, this system would have led to most students leaving school at around age 13 or 14, which was considered too young, so the school admission age was raised from 5 to 6 years so that school leavers would be 15-year-olds and hence more mature. Everyone was taught Mathematics in this program of general education. Religion, first and second Language, Social Studies, integrated Science, Health and Physical education, Aesthetic education, and two prevocational subjects were among other subjects in the common curriculum of about eighty subjects.

In 1972, a constitutional change saw Ceylon become the Republic of Sri Lanka, so the 1972 educational reforms survived less than five years. The General Certificate of Education (GCE), Ordinary Level (OLs), and Advanced Level (ALs) examinations had been replaced by the National Certificate of General Education (NCGE) and the Higher National Certificate of Education (HNCE) respectively, taken at different stages of education. These became very unpopular because of their lack of international recognition; so in 1977 the GCE, OLs and ALs examinations were re-introduced. However, many of the 1972–1977 curriculum reforms were retained up to year 11 in the common curriculum.

The Mathematics syllabus for the HNCE was designed to help a wide range of subjects, like chemistry, physics, biology, psychology, sociology, economics, accounting, commerce and geography. In Grade 11 about 40% of the time was devoted to Mathematics, compared with only about 15% in Grade 10. This syllabus contained some modern topics such as Boolean algebra and matrices. At the GCE and ALs, instead of going back to Pure and Applied Mathematics, the syllabus was split up into two, for engineering and science students respectively. Engineering students had to do the Science syllabus as well. The Science subject included topics like algebra, geometry, complex numbers, calculus, probability and statistics, while the engineering one included differential equations and Newtonian mechanics.

Education Reforms (1997)

The policies for free textbooks, meals, school uniforms and transportation, in place since the 1980s, have made Sri Lanka's education one of the most accessible in the developing world. The next major reform of Mathematics education in Sri Lanka took place in 1997 under the direction of a Presidential Task Force that created the National Education Commission (NEC). The NEC recommended an integrated curriculum, including religion, mother tongue, and Mathematics, using a child-centered approach and environment-related activities, a learning and teaching process involving guided play, desk work and activity, and with assessment throughout the learning and teaching process. This recommendation from the NEC was accepted and by 2003 these reforms had been carried out successfully for both primary and secondary education.

For the key stages 1, 2 and 3 of primary education, Mathematics teachers' manuals indicated 10 common competencies. There was a spiral approach to the syllabus in Mathematics, with depth of knowledge increasing at each stage.

In contrast to the situation in primary education, the proposed 1997 reforms were weak, both implementation-wise and conceptually, at the secondary education level. Students were intended to be sitting GCE or OLs at the end of Grade 11, hence had to be taught methodologically to prepare them for the nationwide exams. Consequently, in 2003, proposals were put forward to improve the quality of Mathematics in junior secondary school. There were then about 2,400 teachers without the training required for this, so teacher development programs were conducted by the National Institute of Education (NIE) and the

Department of Education (DOE) to improve and enhance the skills of junior secondary Mathematics teachers.

For the GCE Advanced Level (Grades 12–13), those students who were taking physical sciences (Chemistry and Physics) were offered combined Mathematics (pure and applied Mathematics). However there were only about 600 schools, i.e., around 25% of all the schools in Sri Lanka, that were offering GCE AL Mathematics. It was proposed by the NEC in 2003 that Mathematics and Science be taught in all the schools (about 1750) across the country which were offering GCE OLs. However, this was not properly implemented and resulted in a decrease in the passing rate among GCE AL Mathematics students.

The structure of paper I of General Certificate of Education Ordinary Level Mathematics is shown in Figure 5.

Part A consists of 30 questions and part B consists of 5 structured questions that must be completed in 2 hours. According to the DOE, in 2012 the GCE OL Mathematics syllabus consisted of Numbers, Algebra, Geometry, Measurements, Statistics, Sets and Probability.

The form of the first General Certificate of Education (GCE) Ordinary Level (OL) paper is given below. Out of the six facility of themes, none of the facilities received more than a 50% pass rate. Statistics had the greatest facility 45% out of the six, while Sets and Probability had the least facility at 31% (see Figure 6).

Students' Achievements in Mathematics

Currently, students' achievements in Mathematics are analyzed mainly in two different stages, at end of Grades 11 and 13, in December every year. GCE OLs are taken by approximately 272,000 students in some 6,600 schools and GCE ALs by approximately 50,000 students in some 700 schools nationwide in Sri Lanka.

* Time : 02 hours. Marks : 50
* This paper consists of two parts, **A** and **B**. All questions included in these two parts are compulsory.

Part A

This part consists of 30 questions which require short answers. It is based on all the themes and covers the entire mathematics syllabus.

1 mark each for questions 1 to 10, with a total of   10 marks.
2 marks each for questions 11 to 30, with a total of 40 marks.
Total marks for Part **A** is   50 marks.

Part B

This part consists of 5 structured questions on the themes Numbers, Measurements, Statistics and, Sets and Probability. These questions are based on the competencies that are closely related to daily life.

10 marks for each question, with a total of 50 marks for Part **B**.

Part **A** -	50 marks
Part **B** -	50 marks
Total marks for Paper I -	100 marks
Final marks for Paper I -	100 + 2 = 50

Figure 5: Source: National Evaluation & Testing Service, Department of Examinations (2012)

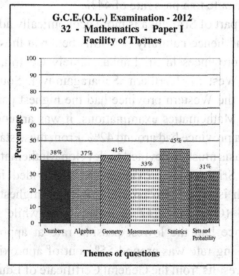

G.C.E.(O.L.) Examination - 2012
32 - Mathematics - Paper I
Facility of Themes

Figure 6. Source: National Evaluation & Testing
Service, Department of Examinations (2012)

Passing rates in GCE OLs Mathematics have increased since 2002 by about 15 points from 40 percent then. Hence, approximately 55% qualify for GCE ALs from GCE OLs even though the mean score in GCE OLs has been unchanged approximately since 2003. The variation in mean scores of GCE OLs in Sri Lanka has been very low, around 32 within 2 points up or down throughout these years, according to the Department of Education's Research and Development Branch (2015).

There are two types of government schools in Sri Lanka, called national and provincial schools respectively. The Ministry of Education manages national schools, while individual provinces manage provincial schools. There is a Grade 5 scholarship exam that would allow provincial school students with high marks to enter the national schools. National schools usually provide better facilities and have better trained teachers and a better learning environment. They perform better in GCE OLs examinations. The pass rate for Mathematics in 2009 GCE OLs in national schools was around 75%, while provincial schools had around 44% pass rate. Because there is only one national school to every four provincial schools, around 20% of students contribute to a pass rate of 75% while about 80% contribute to a pass rate of 44%.

The western part of Sri Lanka is more economically advanced than the eastern parts, and hence naturally performs better in these examinations. There are nine provinces in Sri Lanka, namely Central, Eastern, North Central, North Western, Northern, Sabaragamuwa, Southern, Uva and Western. While the Western province had the highest passing rate in the Ordinary Level Mathematics examinations, it was around 60% in North Central, and Uva province had around 42%. From these statistics, it is clear that more emphasis must be devoted to develop Mathematics in provincial schools. The 9 Sri Lankan provinces are subdivided into 25 districts. Colombo, which is in western province, had the highest passing rate in Ordinary Level (OLs) in 2012 at about 67% while Killinochchi in Northern province had the lowest passing rate at approximately 42%. Overall the passing rate was around 55% out of approximately 270,000 students. Some results from the General Certificate of Education Ordinary Level Mathematics Examination in year 2012 are shown in Figure 7.

While approximately 34,000 students follow the Mathematics curriculum in Grade 13 classes in government schools, almost all of them sit for the combined Mathematics paper for Advanced Level and only a very few sit for the old syllabus Mathematics exams. The passing rate for Advanced Level combined Mathematics steadily decreased after 2004, when it was around 53%, down to about 45% and then increased to 50% in 2012. Figure 8 shows pass rates for provinces in the GCE AL examinations 2015 for all candidates by province. Southern and Northern Provinces had the highest passing rate, 52%, in Physical Sciences, which includes combined Mathematics, Physics and Chemistry. The lowest rate, being 42%, was in North Central region in 2015.

The GCE AL examinations 2015 for all candidates by district are shown in Figure 9. District wise, in 2015, Jaffna in Northern province had the highest passing rate in physical sciences at 56%, while Mullative had the least passing rate at 31%, according to the Department of Examinations.

Standards of Sri Lankan Education

The National Council of Teachers of Mathematics (NCTM) standards for school Mathematics are aligned with Sri Lankan Mathematics standards (McCaul, 2007). The NCTM standards are divided into five process standards: Number and operations, Measurement, Geometry, Algebra, and Data analysis and probability, while Sri Lankan content standards focus on Number, Measurement, Geometry, Algebra, Statistics, Sets, and Probability. Hence, Sri Lanka closely follows the NCTM standards.

According to Mullis et al. (2008), from the Asia Pacific countries, Singapore, the Republic of Korea, Chinese Taipei, Hong Kong and Japan had the highest Mathematics performance among Grade 8 students in years 1999, 2003, and 2007. Singapore was the top performer in fourth and eighth Grades among other countries in 2003. There was a significant gap in Mathematics at Grade 8 Level performance between these five countries and the other countries in 2007.

District	Number Sat	Distinction (A)		Very Good Pass (B)		Credit Pass (C)		Ordinary Pass (S)		Pass (A+B+C+S)		Weak Pass (W)	
		Number	%	Number	%	Number	%	Number	%	Number	%	Number	%
1. Colombo	30431	6161	20.25	3108	10.21	4797	15.76	6329	20.80	20395	67.02	10036	32.98
2. Gampaha	25631	3057	11.93	2079	8.11	3616	14.11	5784	22.57	14536	56.71	11095	43.29
3. Kalutara	14616	1805	12.35	1177	8.05	1987	13.59	3042	20.81	8011	54.81	6605	45.19
4. Kandy	19202	2040	10.62	1435	7.47	2449	12.75	4331	22.55	10255	53.41	8947	46.59
5. Matale	6372	547	8.58	372	5.84	757	11.88	1424	22.35	3100	48.65	3272	51.35
6. Nuwara Eliya	9104	595	6.54	447	4.91	1009	11.08	2129	23.39	4180	45.91	4924	54.09
7. Galle	15013	2088	13.91	1227	8.17	2106	14.03	3214	21.41	8635	57.52	6378	42.48
8. Matara	11285	1478	13.10	909	8.05	1471	13.04	2501	22.16	6359	56.35	4926	43.65
9. Hambantota	8645	884	10.23	612	7.08	1154	13.35	1956	22.63	4606	53.28	4039	46.72
10. Jaffna	8470	1267	14.96	686	8.10	1123	13.26	2039	24.07	5115	60.39	3355	39.61
11. Kilinochchi	1697	86	5.07	67	3.95	185	10.90	367	21.63	705	41.54	992	58.46
12. Mannar	1480	103	6.96	92	6.22	210	14.19	463	31.28	868	58.65	612	41.35
13. Vavuniya	2662	289	10.86	165	6.20	339	12.73	691	25.96	1484	55.75	1178	44.25
14. Mullativu	1407	84	5.97	59	4.19	146	10.38	304	21.61	593	42.15	814	57.85

Figure 7. Regional results. Source: Research & Development Branch, National Evaluation & Testing Service, Department of Examinations (2012. Table 3)

District	Number Sat	Distinction (A)		Very Good Pass (B)		Credit Pass (C)		Ordinary Pass (S)		Pass (A+B+C+S)		Weak Pass (W)	
		Number	%	Number	%	Number	%	Number	%	Number	%	Number	%
15. Batticaloa	6623	620	9.36	453	6.84	1002	15.13	1851	27.95	3926	59.28	2697	40.72
16. Ampara	8784	733	8.34	619	7.05	1245	14.17	2456	27.96	5053	57.53	3731	42.47
17. Trincomalee	5229	351	6.71	283	5.41	574	10.98	1334	25.51	2542	48.61	2687	51.39
18. Kurunegala	21426	2398	11.19	1765	8.24	3149	14.70	5298	24.73	12610	58.85	8816	41.15
19. Puttalam	9395	774	8.24	624	6.64	1228	13.07	2361	25.13	4987	53.08	4408	46.92
20. Anuradhapura	11654	852	7.31	702	6.02	1435	12.31	2586	22.19	5575	47.84	6079	52.16
21. Polonnaruwa	5021	347	6.91	299	5.95	552	10.99	1191	23.72	2389	47.58	2632	52.42
22. Badulla	12400	1022	8.24	802	6.47	1621	13.07	2921	23.56	6366	51.34	6034	48.66
23. Monaragala	6453	371	5.75	331	5.13	690	10.69	1373	21.28	2765	42.85	3688	57.15
24. Ratnapura	14075	1305	9.27	1020	7.25	1839	13.07	3005	21.35	7169	50.93	6906	49.07
25. Kegalle	10783	1182	10.96	794	7.36	1592	14.76	2469	22.90	6037	55.99	4746	44.01
All Island	**267858**	**30439**	**11.36**	**20127**	**7.51**	**36276**	**13.54**	**61419**	**22.93**	**148261**	**55.35**	**119597**	**44.65**

Figure 7 (continued). Regional results. Source: Research & Development Branch, National Evaluation & Testing Service, Department of Examinations (2012. Table 3)

| District | Physical Science | | |
| | Number Sat | Qualified for U.E. | |
		Number	%
1 Southern	5,965	3,084	51.70
2 Northern	1,786	923	51.68
3 Western	10,396	5,229	50.30
4 Sabaragamuwa	2,654	1,314	49.51
5 Central	3,390	1,614	47.61
6 Eastern	1,574	715	45.43
7 Uva	1,711	772	45.12
8 North Western	3,552	1,540	43.36
9 North Central	1,365	574	42.05
Island	**32,393**	**15,765**	**48.67**

Figure 8. GCE AL examinations 2015 for all candidates by province. Source: Department of Examinations (2015)

No.	District	Physical Science Number Sat	Qualified for U.E. Number	Qualified for U.E. %
1	Jaffna	1,227	691	56.32
2	Colombo	5,002	2,671	53.40
3	Matara	2,129	1,132	53.17
4	Batticaloa	423	221	52.25
5	Hambantota	1,564	803	51.34
6	Galle	2,272	1,149	50.57
7	Ratnapura	1,351	683	50.56
8	Kandy	2,214	1,096	49.50
9	Gampaha	3,343	1,646	49.24
10	Kegalle	1,303	631	48.43
11	Trincomalee	302	145	48.01
12	Monaragala	433	199	45.96
13	Vavuniya	183	84	45.90
14	Matale	557	253	45.42
15	Puttalam	747	339	45.38
16	Kilinochchi	119	54	45.38
17	Badulla	1,278	573	44.84
18	Kalutara	2,051	912	44.47
19	Mannar	110	48	43.64
20	Anuradhapura	980	427	43.57
21	Kurunegala	2,805	1,201	42.82
22	Nuwara Eliya	619	265	42.81
23	Ampara	849	349	41.11
24	Polonnaruwa	385	147	38.18
25	Mullaitivu	147	46	31.29
	Island	32,393	15,765	48.67

Figure 9. GCE ALs examinations 2015 for all candidates by district. Source: Department of Examinations (2015)

It is clear that there is a big achievement gap in Mathematics in Sri Lanka across its provinces and districts. Schools have different levels of availability of resources, qualified teachers and teacher training. To improve standards in Sri Lankan school Mathematics, these gaps must be minimized. The UK and Singapore have addressed these issues by providing different Mathematics syllabi to accommodate the needs of different students. Sri Lanka could approach this in a similar manner. It's imperative to increase the knowledge of teachers by making sure they are qualified to teach their subject and giving them adequate training while providing them with the necessary resources. Teachers' instructional manuals should be revised to accommodate these needs better, while continuing to be useful and straightforward. Teachers could also use their own lesson plans and use different strategies to improve the standards of Sri Lankan Mathematics.

Teacher Training

Before the 1980s, government teachers were appointed with Senior School Certificate (SSC), and they were regarded as substitute teachers. The graduates were taken to be teachers but were required to do training in Teacher Training Schools. This training was given to them after they were appointed as teachers. This training included the syllabus, and updating the teachers to the changes in the education system in Sri Lanka. At the end of the 1980s, the training was more efficient.

A 2016 Sri Lanka Government official Gazette mentions "To produce students with full of good thoughts and education, the teacher needs to have good education and good thoughts".

The following institutions train teachers in Sri Lanka at the present:
1. National Colleges of Education (18 Colleges)
2. National Institute of Education
3. Government Universities (colleges)
4. Private institutions

Qualification to become a lecturer in a Teachers College in Sri Lanka are as follow:

1. Post graduate with 1st Class Honors
2. Master of Science or Master of Arts

The way the teachers are recruited in the present times

In 1994, an official "Teacher Profession" was established and the teachers were recruited in two ways:

1. With a College degree/University degree
2. With G.C.E. Advanced Level

To be recruited, teachers must now have College degrees:

1. Have a degree from a reputable Sri Lankan College (University)
2. For the graduates, one year of post graduate diploma training was mandatory from one of the following institutes:
   - From any university (Government colleges)
   - Open University of Sri Lanka
   - National Institute of Education
3. The teachers are given training in the following areas:
   - The Sri Lankan education structure
   - The new education structure
   - Practical education in cognitive science.

Teachers may also be recruited with GCE Advanced Level. Students with excellent results on GCE Advanced Level that are close to entering Sri Lankan Universities (Colleges) may take a competitive exam, focused on measuring knowledge and the logical thinking that is needed for a teacher. These students are sent to one of 18 National Colleges of Education. In National colleges, three years of training are mandatory, including two years of residential training and one year of internship training in schools (teaching in schools under supervision).

The first two years of the training focuses on the following:

1. Subject matter
2. Educational thinking
3. Techniques of teaching

Other training opportunities include workshops that are held when there is a change in the syllabus, when there are any changes in the laws, or other situations like these when teachers need to be educated with updated information.

The north and east after the war

With the above procedure, the teacher training in this area is done using their language. Most of the people use Tamil as their language compared to Sinhalese in the rest of the country. The war-affected areas are being reconstructed (damaged buildings, scientific equipment, etc.), according to the needs, by the government. The leaders of these areas, like government ministers, are elected, and hence they know the needs of the local people.

References

Little, A. W. (2010). *The Politics, Policies and Progress of Basic Education in Sri Lanka*. Data Management Branch, Ministry of Education Sri Lanka, Department of Census and Statistics, Ministry of Education and Central Bank of Sri Lanka.

Department of Education (2006). *GCE OL Examination, Controlling chief examiners' meeting report*. Department of Education, Sri Lanka.

Department of Education (2008). GCE O.L. *Examination, Controlling chief examiners' meeting report*. Department of Education, Sri Lanka.

Department of Education (2009). *GCE O.L. Examination, Controlling chief examiners' meeting report*. Department of Education, Sri Lanka.

Department of Education (2010). *GCE O.L. Examination, Controlling chief examiners meeting report*. Department of Education, Sri Lanka.

Department of Education (2012). *GCE O.L. Evaluation Reports*. Research & Development Branch, National Evaluation & Testing Service, Department of Examinations, Department of Education, Sri Lanka.

Department of Education (2015). GCE A.L. *Performance of Candidates, Evaluation Reports*. Research & Development Branch, National Evaluation & Testing Service, Department of Examinations, Department of Education, Sri Lanka.

LankaWeb (No date). Dr. C.W.W. Kannangara: Father of Free Education. http://lankaweb.com

McCaul, T. (2007). *A study of the Implementation of Mathematics and Science Curriculum in Grade 6 and 10*. National Institute of Education, Sri Lanka. Ministry of Education, http://www.more.gov.lk

Mullis et al. (2008). *TIMSS 2007 International Mathematics Report: Findings from IEA's Trends in International Mathematics and Science Study at the Fourth and Eighth Grades*. National Education Commission, Sri Lanka.

NCTM (2010). *Principles and Standards.* National Council of Teachers of Mathematics. http://www.nctm.org/standards

National Institute of Education (2006). *Report Submitted to the Director General of the NIE by the Committee for the subject of Mathematics*. National Institute of Education, Sri Lanka.

National Institute of Education (2008). *Primary and Junior Secondary Syllabus.* National Institute of Education. http://www.nie.lk/page/syllabus.html

National Education Commission (2003). *Proposals for a National Policy Framework on General Education in Sri Lanka.* National Education Commission, Sri Lanka

SLAAED (2000). Education Reforms in Sri Lanka, Sri Lanka Association for the Advancement of Education.

South Asia Human Development Sector (2011). *Strengthening Mathematics Education in Sri Lanka.*

The Institution of Engineers of Sri Lanka (no date). *The Institution of Engineers of Sri Lanka: Home page.* http:// www.iesl.lk

About the Author

Mr. Lucian Makalanda has been teaching in City University of New York (CUNY) Queensborough Community College for close to ten years. He also has an initial certification of teaching secondary education from Queens College. He attended the Sri Lankan school system for his primary and secondary education in Mathematics and graduated from University of Sri Jayewardenepura in 2003 with a Bachelor of Science in Mathematics, Statistics and Physics (Applied Science). He also has a Masters in Mathematics from Queens College (CUNY) and is currently working towards his Doctor of Education in College Teaching (Ed.D.C.T.) at Teachers College Columbia University. Mr. Makalanda is also the coordinator of Statistics in Queensborough Community College (CUNY).

Printed in the United States
by Bookmasters

Printed in the United States
By Bookmasters